Seeing Our Planet Whole: A Cultural and Ethical View of Earth Observation

Harry Eyres

Seeing Our Planet Whole: A Cultural and Ethical View of Earth Observation

 Springer

Harry Eyres
European Space Policy Institute
Vienna, Austria

ISBN 978-3-319-40602-2   ISBN 978-3-319-40603-9   (eBook)
DOI 10.1007/978-3-319-40603-9

Library of Congress Control Number: 2016947051

Printed on acid-free paper

This Springer imprint is published by Springer Nature
The registered company is Springer International Publishing AG Switzerland

To Peter Hulsroj and his team at ESPI—a fine example of cooperation in wide-ranging intellectual endeavour

Acknowledgements

As well as the excellent institutional support provided by ESPI, I would like to acknowledge the invaluable help offered by ESA and in particular Josef Aschbacher at ESRIN. Wolfgang Rathgeber, Mark Doherty, Francesco Sarti and Pierre-Philippe Mathieu were supportive guides during my time at ESRIN. At EUMETSAT, I would like to thank Alain Ratier, Marc Cohen and Paul Counet. My work has been greatly strengthened and enriched by numerous conversations with Herbert Allgeier and Roy Gibson. Intellectual stimulation has come in unstinting measure from Peter Hulsroj and also from Marco Aliberti, Arne Lahcen and Stefano Ferretti. Peter Hulsroj, Josef Aschbacher, Roy Gibson, Herbert Allgeier, Marco Aliberti, Arne Lahcen and Stefano Ferretti read the manuscript at various stages and I have tried to incorporate their many helpful comments. Any errors and oversights which remain are down to me. Stefano Ferretti and Cenan Al-Ekabi have also assisted me on many practical matters. Ching Ling has been, as always, a great support.

Contents

1

Introduction

Italo Calvino's short novel The Baron in the Trees, set in Liguria in the eighteenth century, is the tale of a boy, Cosimo, who climbs a tree after an argument with his sister and decides never to come down. He manages to live a self-sufficient life in the treetops. The philosopher Voltaire hears of his exploits and asks Cosimo's brother Biagio whether Cosimo went to live in the treetops in order to be nearer to the sky. Biagio replies: "My brother considers that anyone who wants to see the earth properly must keep himself at a necessary distance from it."[1]

Cosimo's act of distancing himself from ordinary society is not a retreat from it. On the contrary, from his arboreal standpoint he is passionately concerned with helping and reforming his fellow humans. As his brother Biagio comments, "Cosimo lived almost as closely with us as he had before"[2] and became "at the same time a friend to his neighbour, to nature and to himself."[3]

Cosimo can be seen as a pioneer of Earth Observation, and of its revolutionary potential to be a friend to both nature and mankind. Earth Observation in turn can be seen as one of a number of scientific and philosophical tools discovered in the last century and a half which have enabled human beings to observe themselves and their environment from a "necessary distance". Others are evolutionary biology, genetics and psychoanalysis.

---

[1] Italo Calvino, *The Baron in the Trees* (London: Collins, 1959) p. 170.
[2] Ibid, p. 78.
[3] Ibid, p. 122.

© Springer International Publishing Switzerland 2017
H. Eyres, *Seeing Our Planet Whole: A Cultural and Ethical View of Earth Observation*, DOI 10.1007/978-3-319-40603-9_1

Know Thyself

Earth Observation, together with the various natural and social scientific disciplines, can be considered part of an ongoing project of knowing and understanding ourselves and our environment. This project was initiated in the West in two words inscribed above the entrance to the temple of Apollo at Delphi in ancient Greece, gnothi seauton or know thyself.

If such a project were easy, there would be no need to be reminded of it. The earliest Western philosopher of whom we have an extensive record, Socrates, took this project to heart. Initially, he tells us, he was interested in what we would now call natural scientific inquiry, how the world worked, what it was composed of and so on. But this always led to a still greater mystery, the workings of the human mind and soul. Socrates decided to leave his "scientific" enquiries on hold and concentrate on metaphysics and psychology.[4]

Socrates's misgivings about natural scientific enquiry may seem misguided considering the enormous achievements of western science especially in the last 400 years. However, despite our great advances, equally vast areas of ignorance remain. As Galileo said, the small amount of understanding he might have achieved only made him consider the immense amount he did not know. Galileo's humility chimes with the recognition by contemporary astrophysicists that close to 95 % of the universe, consisting of dark matter and dark energy, remains completely unknown.

The example of Galileo may also serve to remind us of the obfuscatory power of human institutions and ideologies. The Catholic Church took over 350 years to recognise the validity of Galileo's observations confirming the Copernican heliocentric theory. The seventeenth century Archbishop James Ussher's dating of the creation to 4004 BC was upheld by many theologians until the middle of the nineteenth century.

But before we pride ourselves on our superior knowledge, we might pause to wonder whether we ourselves could be wearing comparable ideological blinkers. For the last four hundred years or so, we in the West at least have been under the influence of the new kind of natural philosophy announced by Francis Bacon, opening the way to the scientific revolution. This would dethrone ancient authority, especially the Bible and the work of Aristotle, as the basis of knowledge and replace it with empirical observation. At the same time it would set no limits on the exploitation of nature for human ends. We may be reaching a point where the two prongs of the Baconian experiment are twisting apart: science itself is telling us that the unbridled exploitation of the natural world cannot be continued ad infinitum.

---

[4] See Plato, *Phaedo*.

Observing the Earth as a Totality

People of course have always observed aspects of their immediate and more distant environment; we evolved from forest-dwelling apes into cave-dwelling humans not without close observation of our surroundings—of the plants which might nourish or harm us, of predatory animals which might prey on us and others on which we might prey, of the changes of the seasons and the vagaries of the weather. From earliest times no doubt human beings also looked up into the sky, observed and wondered at the heavenly bodies.

Earth Observation, however, refers to the ability to view the earth as a totality, a single living globe, the only heavenly body so far discovered which can sustain life; our home planet: our single almost certainly irreplaceable home. In this sense Earth Observation is a very recent phenomenon, dating back to the launch of the first satellites with observational capacities in the 1960s.

As Italo Calvino's beautiful tale implies, when we are too close to anything (as are most of the inhabitants of Liguria, immersed in their petty squabbles), we cannot observe it properly; as the popular saying goes, we cannot see the wood for the trees. Likewise when we are too far away, too much detail is lost; the detachment is too great. Earth Observation provides us with the necessary distance to get a clearer view of things, to see the Earth as a whole, and to assess changes on local, regional and planetary scales, from shifting patterns of vegetation in one particular woodland to deforestation in tropical rain forests, the melting of glaciers and icecaps and changes in the level and temperature of the oceans.

The relatively new science of ecology teaches us that everything is connected. Life on earth is a set of intimately and intricately interconnected relationships. Human beings have always known they depended on clean air to breathe, water to drink, a climate neither too icy nor too torrid, food to forage, hunt or grow. We are only just beginning to become aware of the complex interdependence of species and habitats. For instance, scientists studying sperm whales have begun to understand the very important role vast cetaceans play in "recycling" small marine organisms from deeper to shallower waters. This is no doubt only one of innumerable such interrelationships; it is also one which can be monitored (in terms of whale populations) using satellites.

The Ecological Crisis

For decades now, scientists have been warning of an ecological crisis, caused by our own impacts and actions. The crisis is complex and multi-dimensional.

In the early 1960s the biologist and journalist Rachel Carson sounded the alarm about the overuse of chemical pesticides. Some of the products have been banned, but the poisoning of the earth, oceans and atmosphere continues. Levels of DDT found in Adelie penguins in Antarctica are the same as thirty years ago.[5]

From the 1970s onwards, the signs became manifest that human greenhouse gas emissions were warming the climate. The vast body of scientific research done since then, and collated by the IPCC in its reports, has confirmed the initial fears. The actual and potential effects of climate change, ranging from sea level rise, partly caused by the melting of the Greenland and Antarctic ice sheets, to glacier melt, to increased risks of drought and extreme weather events, are extremely wide-ranging and potentially devastating.

Other atmosphere-borne threats, apart from the build-up of greenhouse gases, have been apparent since the 1970s—acid rain and damage to the ozone layer. The increasing concentrations of CO_2 in the atmosphere were found, especially from the early 2000s, to be reflected by the acidification of the oceans.

Apart from these fairly drastic threats to the well-being of humans and other species, there has been the unrelenting pressure of increased human population and impact on the natural world: deforestation in tropical regions (itself a major and double cause of increasing CO_2 emissions, because the burning of forests releases CO_2 while the reduction of forest cover reduces the earth's capacity to absorb CO_2), the destruction of habitats and the most extreme cases the extinction of species. A number of eminent biologists including the Harvard entomologist E.O. Wilson consider that we have embarked on the sixth great extinction event in the known history of the earth, and the only one caused by our own actions.

The utility of satellite Earth Observation in monitoring such threats to the environment, as well as others such as deforestation and marine pollution, is perhaps reasonably well-known. In relation to climate change, 26 out of 50 "essential climate variables" can only be monitored from space. In May 2015 the European Space Agency's Cryosat mission detected a sudden loss of ice in the southern Antarctic peninsula, a previously stable zone.[6]

In this study we will consider Earth Observation from less familiar angles—from cultural, ethical, aesthetic and democratic perspectives as well as the more familiar scientific and economic ones. In so doing we will ask whether

---

[5] http://pubs.acs.org/doi/pdf/10.1021/es087118g
[6] http://www.esa.int/Our_Activities/Observing_the_Earth/CryoSat/CryoSat_detects_sudden_ice_loss_in_Southern_Antarctic_Peninsula

this apparently techno-scientific tool has even wider applications than we have so far imagined.

In particular we propose that Earth Observation can be envisioned and used as an ethical and cultural tool or resource, not just a techno-scientific one. This may seem counterintuitive, or even perverse. Surely Earth Observation, the product of the highest levels of science and engineering, is scientific and technical by nature. Here a famously gnomic remark by Heidegger in his essay The Question Concerning Technology might give us pause. "The essence of technology is by no means anything technological."[7] In the end Heidegger suggests that the essence of technology resides in poesis or bringing forth into being, not as a means to an end.

The essence of Earth Observation is not merely to gather data, still less to efficiently manage and exploit the earth. No, the essence of Earth Observation is ethical and connected to responsibility and care. It is linked also to basic human motivations, instincts and feelings: curiosity, wonder, awe. Here we indicate a momentous shift from the paradigm which has been dominant since the beginning of the scientific revolution in the early seventeenth century, based on Francis Bacon's granting of licence to twist and torture mother nature for exclusively human ends, and sanctioned by the abstraction of soul and feeling from a mechanical universe.

We are understandably reluctant to let go of a paradigm which has apparently served humanity well. The wealth and comfort of the majority of people have increased immeasurably since the time of the scientific, and more particularly the industrial, revolutions. But the warning signs that ever-increasing human numbers and prosperity are putting dangerous strains on the carrying capacity of the planet have been flashing for some time; the blithe assumption that the earth can cope with all our pressures, including our emissions of greenhouse gases, is no longer sustainable. Earth Observation contributes crucially to global capacities of environmental monitoring, with special relevance to indicators of changing climate. At the same time the evidence it provides is essentially unarguable; it is not a matter of theories or projections but of incontrovertible data.

The Delphic command to know ourselves must be extended to the environment, as we cannot exist without it. Knowing ourselves means knowing our own responsibility, especially for any long-term damage or harm. Earth Observation is an indispensable tool in this project of self-knowledge.

---

[7] Martin Heidegger, *Basic Writings* ed. David Farrell Krell (Routledge, London: 1993) p. 311.

2

Cosmology and Astronomy
from Prehistory to the Roman Empire

In all pre-modern cultures, the great powers of nature, sun and moon, life and death, winds, storms, floods, earthquakes, volcanos, were thought of in divine terms. These great powers were gods, with terrifying and arbitrary potentials, needing to be appeased, often by sacrifices (which in some cultures such as the Aztec one took human form). Human beings seemed small and weak compared to such elemental and often destructive forces. The idea of human beings coming to dominate most of the planet, apart from the deep oceans and polar regions, and even being a threat to its eco-systems, would have seemed far-fetched.

Some of the earliest human cultural manifestations are cave-paintings, found in all continents of the world and dating back 30,000–40,000 years. Many of the paintings depict animals, such as bison, horses, aurochs and deer, with extraordinary grace, beauty and energy, and attest to a very close relationship between humans and animals. One theory is that the paintings were made by shamans, priest-like figures who were regarded as intermediaries between the human world and the spirit world.

From very early times, human beings have speculated about the shape and size of the earth and its relationship to other visible heavenly bodies, especially the sun, moon and planets. These inquiries were no doubt stimulated by obvious phenomena such as the apparent disappearance and reappearance of the sun, the waxing and waning of the moon, the ebbing and flowing of tides, the progression of the seasons, and more dangerous events such as storms, volcanic eruptions and earthquakes. In the chapters which follow, we trace the evolution of such thinking, from the momentous breakthroughs of early

© Springer International Publishing Switzerland 2017
H. Eyres, *Seeing Our Planet Whole: A Cultural and Ethical View of Earth Observation*, DOI 10.1007/978-3-319-40603-9_2

Babylonian, Greek, Hellenistic and Arabic astronomers, through to the revolutionary discoveries of Copernicus, Galileo, Kepler and Newton.

Even though fear plays a indubitable part in the attitude of ancient cultures to what we now call environment, traditional cosmogonies—accounts of the origins of the universe—and cosmologies stress the order and beauty of the cosmos. Indeed the Greek word cosmos also means order.

The Timaeus and the Old Testament

Perhaps the most influential of Greek cosmogonies comes in Plato's dialogue Timaeus. The main speaker, Timaeus, says that "the creator was good, and... desired that all things should be as like himself as they could be...Out of disorder be brought order."[1] For Timaeus the world is not only good and ordered, but shares the characteristics of a living being: "the world became a living creature truly endowed with soul and intelligence by the providence of God."[2]

Choosing the shape the world would take, God "made the world in the form of a globe...the most perfect and the most like itself of all figures."[3] The idea of the earth as a sphere, according to Diogenes Laertius, was first proposed by Pythagoras. It was confirmed by the work of Hellenistic astronomers, especially Eratosthenes, in the third century BC. This globed perfection would be finally appreciated when it was possible to see the earth from space.

The beauty, order and living perfection of the created world give pleasure to the creator: "when the father creator saw the creature which he had made moving and living, the created image of the eternal gods, he rejoiced."[4]

This rejoicing is echoed by the creator in the Biblical book of Genesis. After each of the first six days of creation, when God creates day and night, the earth, the seas and the heavens, the sun, the moon and the stars, and then the living creatures that populate land, sea and sky, culminating with man himself, he tells himself that "it was good", or after the sixth and final day of creation, that "it was very good."

In Genesis God commands Adam and Eve, the forefather and mother of all mankind, "to replenish the earth, and subdue it; and have dominion over the fish of the sea, and over the fowl of the air, and over every living thing that

---

[1] Plato, Timaeus, tr. Benjamin Jowett http://classics.mit.edu/Plato/timaeus.html
[2] Ibid.
[3] Ibid.
[4] Ibid.

moveth upon the earth." But the command to subdue and have dominion over the earth and other species does not mean to destroy it and exterminate them. Even when God is so disgusted with his creation that he repents having made it, he stops short of extermination: he sends the Flood to destroy the earth and its creatures, but preserves Noah, because of his uprightness, and then commands him to build his ark and fill it with two of every species on earth, so as to "keep seed alive upon the face of the earth."

The Flood lasts for a hundred and fifty days, drowning everything on earth, so that only Noah, his family and the animals on the Ark are saved. Then God remembers Noah and the animals with him, and causes the waters to be "assuaged". The ark finally reaches dry land and God establishes his covenant with Noah and with the animals he has taken on board. From this time "neither shall all flesh be cut off…neither shall there be a flood to destroy the earth."

These traditional kinds of cosmogony and cosmological thinking, whether Greek or Hebrew, bring with them certain ethical attitudes towards what we now call environment. Whatever the differences between them, they share ideas of a beautiful natural order perfected by god, that is beyond and before human agency, with humanity in a privileged but not all-controlling position.

In the Old Testament, perhaps the most profound thoughts about the relationship between mankind and the natural order come in the book of Job. Job is a righteous man whom God allows to be tested by Satan, who wants to prove that his apparent piety is skin-deep. Job loses his wealth, his children and his health. Three so-called friends or counsellors come to offer smug explanations, maintaining that Job's sufferings are the just punishment for sin. Job indignantly rejects them. A fourth friend, Elihu, speaks to Job of the unfathomable wisdom of God.

Eventually God appears, speaking to Job from a whirlwind. He does not answer any of Job's specific questions and complaints about the injustice of his sufferings. Instead, in some of the most profound poetry of the Old Testament, he takes Job on a tour of the deepest recesses and mysteries of creation, asking him "where wast thou when I laid the foundations of the earth?"

The purpose of God's speech is to set Job's limited human perspective against the infinite grandeur of the universe, with its processes which surpass human knowledge, understanding and control.

By the end, Job confesses his own ignorance of "things beyond me which I did not know." God restores to him his lost wealth and health. The lesson of the Book of Job may appear far removed from present concerns, but it is also possible to see Earth Observation as a way of restoring our sense of wonder and awe at the planet's beauty and complexity.

Greek Philosophy, Religion, Tragedy

Roughly contemporary with Job (700–400 B.C.) are the first developments of Greek science which would eventually lead humanity in a very different direction. Quasi-scientific enquiries into the natural world began in Ionia (present-day western Turkey) in the sixth century B.C. According to Lloyd and Sivin, "[The Greeks] invented the concept of nature to serve distinct polemical purposes—to define their sphere of competence as new-style investigators and to underline the superiority of naturalistic views to the traditional beliefs of poets, wise men and religious leaders."[5] Philosophers such as Thales of Miletus, Anaximander, Anaximenes and Heraclitus sought materialistic explanations for the origins and composition of the universe. Thales believed the underlying essence of everything was water while Anaximenes attributed that role to air. Heraclitus stressed the dynamic, changing nature of all reality with his aphorism panta rhei (everything flows), and also gave primary importance to fire. His view was opposed by Parmenides, who argued, in terms which would profoundly influence Plato, that shifting appearances masked the essentially unchanging nature of reality. Empedocles combined some of the ideas of the Ionian philosophers with his four-part classification, Earth, Water, Fire, Air, which proved remarkably durable. Meanwhile the atomists Democritus and Leucippus posited a division into atoms (the basic matter of the universe, in its smallest form) and void.

The limitations of materialism, and the question of what gives form to all matter, were highlighted by Pythagoras's view of number as the underlying principle of the universe. It was the Pythagorean view, filtered through Plato and then indirectly into Christianity, which would have the greater influence at least in the short term than the materialist theories; these remained as a strong undercurrent in classical thinking, resurfacing repeatedly, for instance in Epicureanism especially as interpreted by the Roman poet Lucretius in his great long didactic poem De Rerum Natura (On the Nature of Things).

There is obviously already a tension in the ancient Greek view of the relationship between man and the cosmos, though this tension remains largely beneath the surface. This tension is perhaps best expressed in the myth of Prometheus. The elements of the story are given in two accounts by Hesiod and Aeschylus. The titan Prometheus is punished by Zeus for bringing or restoring fire, monopoly of the gods, to mankind. The scene at the start of Aeschylus's

---

[5] Geoffrey Lloyd and Nathan Sivin, The Way and the Word: Science and Medicine in Early China and Greece (New Haven and London: Yale University Press, 2002), p. 241.

tragedy *Prometheus Bound* is unforgettably dramatic: the titan chained to the rock in the Caucasus, forced to have his liver repeatedly devoured by an eagle. Already Prometheus is an ambivalent figure, accused in Hesiod of bringing what we would now call industrial pollution to an idyllic rural economy, in Aeschylus object of sympathy as man's helper and champion.

Prometheus is a disruptive, transgressive figure, symbolic of mankind's technological ingenuity and desire to better its lot, rather than remain within the limits prescribed by the gods. There is a parallel here with the figure of Satan in the Book of Genesis, who persuades Eve to disobey God's commandment and eat the apple, the fruit of the tree of the knowledge of good and evil. Prometheus will resurface again to dramatic effect in later European thinking, starting with the work of Francis Bacon at the beginning of the scientific revolution.

The Hesiodic view of Prometheus as dangerous transgressor probably had more influence on mainstream Greek thinking. This thinking was predominantly religious. Greek religion was polytheistic rather than monotheistic. Homer portrays the Olympian gods as a badly behaved family, full of feuds, jealousies and revenges. Beneath the level of the Olympians, the Greeks worshipped many minor deities, often associated with places or features of the natural world: the water nymphs called Naiads, the mountain nymphs or Oreads, the god of bees and bee-keeping, Aristaeus, Corymbus the god of the fruit of the ivy and many others.

Among the Greek gods we might single out as especially relevant to our discussion the earth goddess Ge or Gaia (later to give her name to James Lovelock's theory of the earth as a self-stabilising organic entity) and Demeter. Gaia is the original Mother of Creation. In Hesiod's Theogony Gaia emerges from Chaos to be the seat of the Immortals, then creates Ouranos (Sky) to be her equal, covering her on every side. From the union of Gaia and Ouranos come the hills, the sea, and the Titans, including Cronos, the father of Zeus who will become King of the Olympians. Thus the Earth, genealogically, has precedence over the King of the Gods.

Demeter, though younger, is a more important deity in terms of cults and mysteries devoted to her in classical Greek times. She is a granddaughter of Gaia and sister of Zeus. Demeter is the goddess of fertility and the harvest; she also presides over the sacred law (thesmos) and the cycles of life and death. In one of the most haunting Greek myths, Demeter's virgin daughter Persephone is abducted by Hades, lord of the Underworld. Demeter searches everywhere for her, distracted by grief, and the earth, in sympathy, ceases to bear fruit. Eventually Hades agrees to release Persephone, but gives her a pomegranate; by eating the seeds, Persephone binds herself to be with Hades for three months of the year, usually interpreted as the fruitless months of winter.

Here we can invoke the familiar notions of hubris and nemesis, especially as explored in Greek tragedy. Some modern scholarship has cast doubt on the extent to which hubris should be seen in a religious context, and stressed its normal meaning in Greek law as insulting behaviour usually of a sexual nature. However, Greek tragedy and especially the plays of Sophocles have often been interpreted in terms of hubris, that is arrogant, overbearing or overweening behaviour, being punished by nemesis, or divine retribution.

In the *Antigone*, for instance, the tyrant Creon issues an edict forbidding the burial, according to traditional religious ritual, of the body of Polynices, one of the sons of Oedipus and one of the leaders of the seven armies attempting to overthrow Thebes. Polynices' sister and Creon's niece Antigone disobeys the edict, buries her brother and is punished by being walled up alive in a cave by Creon. Confronting Creon, Antigone argues that his edict was a piece of arbitrary and tyrannical human law, not underpinned by the deeper divine jurisdiction of Zeus and Dike (justice). Retribution comes to Creon when his son Haemon, betrothed to Antigone, falls on his sword after the suicide of Antigone. Creon's woes are compounded by the suicide of his wife. Seeing his wife's body, Creon cries out "this is my guilt, all mine."

Earlier, in the famous "Polla ta deina" chorus, the chorus celebrates the ingenuity of man ("there are many wonders, but nothing more wonderful than man"), who has been able to tame horses, protect himself from the elements, develop the powers of thought and language, construct cities, but also warns against the way this extraordinary power can be used for good or ill.

There are implications here for a thinking about the relationship between man and nature. Indeed the Greek ideas of hubris and nemesis have sometimes been invoked by contemporary environmentalists, in the thought that human arrogance, in the form of unrestrained exploitation and pollution of the planet, is being repaid by the nemesis of global warming and the poisoning of the planet.

Greek philosophy is in general intensely preoccupied with such subjects as politics, ethics and epistemology, and shows rather less interest in the non-human world.

A partial exception comes in the work of the most influential Greek philosopher, Aristotle. Aristotle can be seen as pointing in at least two different directions. On the one hand (in the *Nichomachean Ethics*) he sees man as part of the natural world, which (following Plato) is by nature good and whose balance needs to be maintained. On the other, in the *Politics*, Aristotle initiates what contemporary environmentalists call the anthropocentric view, when he states in that "nature has made all things specifically for the sake of man."[6]

---

[6] Aristotle, *Politics* Book 1, Ch. 8.

Aristotle is the also the philosopher who initiates the study of natural history, with his *History of Animals* which contains remarkably accurate descriptions of creatures such as the chameleon and observations of the migration of fish.

Hellenistic and Babylonian Astronomy

The great advances of Hellenistic astronomy rested to a considerable extent on the achievements of much earlier Bablyonian and Chaldean astronomers. The Babylonians were the first to recognize the periodicity of the appearance of planets as observed from the earth. The Venus tablet of Ammusduqa, whose origins probably date back to the seventeenth century B.C., records times of the rising and setting of Venus over a 21-year period. Other Babylonian discoveries include eclipse cycles and catalogues of stars. Ancient Greek and Hellenistic astronomers seem to have had access to Babylonian texts and records and to have used them in making their calculations, though we have no direct evidence of this.

Among influential Hellenistic astronomers, Eratosthenes calculated the size of the earth with remarkable accuracy. Most strikingly, perhaps, Aristarchus of Samos put forward a heliocentric model of the universe, and hypothesized that the sun and the moon were much larger than previously supposed, and that the stars were unmoving distant suns. It was not until Copernicus 1800 years later that another astronomer dared to revive these heretical ideas. The most celebrated Hellenistic astronomer, Hipparchus, made very accurate calculations of the lengths of the lunar month and the solar year—although he dismissed Aristarchus' heliocentrism. These findings, including the Babylonian ones, but excluding Aristarchus' heliocentrism, were summed up in Ptolemy of Alexandria's Almagest, which includes a catalogue of over 1000 stars and which dominated astronomical thinking in the West for 1200 years.

Rome

The official religion of the Roman world, up until the coming of Christianity, followed the ancient Greek Olympian pantheon, renaming the gods in Roman style, so that Zeus became Jupiter, Hera Juno, Athena Minerva, Ares Mars and so on. But the Romans maintained certain vestiges of local Italian cults or religions, especially in the form of the small, modest household gods, the lares and penates, whose shrines can be seen in the remains of houses in Pompeii and Herculaneum.

The Roman poets, especially Catullus, Virgil and Horace, are especially appreciative of the beauty of particular, rather than mythological places. They could be seen to initiate the "landscape" tradition in Western art and literature which is an important strand in environmentalism, starting with such figures as Wordsworth, Thoreau and John Muir.

Virgil's second major work, The Georgics, is a long didactic poem on agriculture, divided into four books which include detailed descriptions and advice on weather, ploughing, viticulture, animal husbandry and beekeeping. The Georgics is also notable for its lyrical digressions on the beauty of the Italian landscape and the coming of spring, both in Book Two. But the tone of the Georgics is by no means that of pastoral idyll: a long description of a devastating storm breaks into Book 1, and in Book 3 there is an account of an outbreak of pestilence which kills flocks and herds in Noricum.

For all that, The Georgics is a vision of an essentially harmonious relationship between man and the natural world mediated through what we might now call a sustainable agriculture. Perhaps even more profoundly, this is a poem about peace. In the great concluding passage of Book 1 Virgil looks forward to the time when farmers ploughing the battlefield of Philippi, where thousands of Roman died fighting each other in the biggest land battle of the Civil War, will turn up rusting javelins and empty helmets. Here is a symbol of peaceful agriculture triumphing over war.

Virgil's friend and near-contemporary Horace fought against Antony and Octavian at Philippi but ended up, like Virgil, as a supporter of the Augustan regime which he saw as bringing relative peace after decades of civil war. Throughout his poetic oeuvre, Horace seems torn between the attractions of the city and country.

Horace, who lived part of the time in the West's first mega-city, Rome, at a time of rapid urban expansion under the first Roman emperor Augustus, is perhaps the first writer to complain about urban pollution in recognizably modern terms. In one of his greatest odes, Horace advises his patron Maecenas to leave behind "the smoke and wealth and din of Rome".[7] In this typically compressed phrase is condensed much of the logic of contemporary environmentalism: the link between excessive wealth and pollution.

In another of his most deeply felt and prophetic poems, Horace comes up with an image of still more astonishing modernity. Referring to the building of concrete fishpens, Horace writes that "the fish note the narrowing of the waters by piers of rocks laid in their depths."[8] The Romans were pioneers

---

[7] Horace, *Odes* 3.29.
[8] Horace, *Odes* 3.1.

of pisciculture or fish-farming; what is extraordinary in Horace's lines is the sudden shift to the point of view of the fish as they feel their environment contracting.

Though admitting to pulls in both directions, Horace in the end opted pretty firmly for the healthier balance of life in the country, looking after the farm in the Sabine Hills given to him by Maecenas, over the febrile attractions of the city. Here once again he was prophetic, prefiguring the ruralist strand in Western Green thinking which one can trace through Ben Jonson's To Penshurst, Wordsworth's nature poetry through to Thoreau's Walden (preferring existence in a solitary lakeside cabin to the life of "quiet desperation" in the town).

The nostalgic element in Horace's thinking, the tendency to look back to a golden age of frugality and harmony, is echoed by other Roman writers. At the start of the Metamorphoses, for example, Ovid charts the development, and descent, of man from the uncorrupted Golden Age of perpetual spring, unmarked by agriculture, war or civil discord, through the Silver, Bronze and Iron ages, when the seasons, the need to labour and build shelter, and eventually wars and general deceitfulness and envy spread their miasma over the earth.

3

Aquinas to Newton

The idea of the Dark Ages extending from the fall of the Roman Empire well into the Christian Middle Ages is no longer fashionable among scholars. Among other things, it neglects the persistence of the Eastern Roman Empire based in Constantinople (Byzantium) for a thousand years after the fall of Rome. This period also saw the great flowering of Islamic civilisation, mathematics, science and culture, partly based in what is now Spain, under the Abbasid caliphate. Celtic culture also flourished during this period.

A key text for understanding the mainstream high medieval and early Renaissance Christian position on the relation of humans to the environment is St Thomas Aquinas's Summa Theologiae. This is the greatest work of medieval Christian theology, and, as its name suggests, a summation and reconciliaton of classical, especially Aristotelian philosophy and scholastic theology.

Aquinas

At first glance, Aquinas's positions in the Summa appear both hierarchical and anthropocentric. Following the dominant Biblical and Classical traditions he proposes a hierarchy of creation in which humans occupy a position near the top. Furthermore, the quality of humans which justifies this high position is rationality, which would seem to leave non-human beings devoid of intrinsic value.

© Springer International Publishing Switzerland 2017
H. Eyres, *Seeing Our Planet Whole: A Cultural and Ethical View of Earth Observation*, DOI 10.1007/978-3-319-40603-9_3

But a closer reading of Aquinas, the scholar Martin J. Tracey has argued, reveals "that far from being indifferent or hostile to the environment, Thomas is deeply respectful of it".[1]

Though Aquinas follows the Old Testament line that man has "dominion" over other creatures, he also holds that all creatures and aspects of the natural world were created by God, and have in some sense a divine purpose. For instance, commenting on Psalms 146.9 Aquinas notes that the ravens mentioned by the Psalmist "are said to call upon God on account of the natural desire whereby all things, each in its own way, desire to attain the divine goodness." If ravens as well as humans have divine longings, they must be accorded respect.

Another Thomian text, less frequently cited than the Summa, is the short and incomplete treatise De Regno ad Regem Cypri, written for the King of Cyprus in 1265. In this Aquinas sets down five criteria for the siting of a good city: it should have a temperate climate, wholesome air, clean water, fertile soil and a setting of natural beauty. Here are the bases of an environmental policy and ethics which might seem quite modern. The emphasis on wholesome air—important because of the intimacy of the act whereby we "draw [air] down into our very vitals"—and clean water seems prescient, but the most interesting criterion is the last one. According to Aquinas, the good city "must charm [its] inhabitants by its beauty". He lists the aspects of beauty which are important for a suitable site: there must be "a broad expanse of meadows, an abundant forest growth, mountains to be seen close at hand, pleasant groves and a copiousness of water."[2] Here again Aquinas seems to be ahead of his time in stressing the importance of environment for human flourishing.

The Great Chain of Being

Just as important as the precise theological positions reached by Aquinas in the Summa is the more widely diffused idea, highly influential from medieval times right down to the eighteenth century, of a "Great Chain of Being".

One of the most famous articulations of this idea comes in Ulysses' great speech on Degree in Shakespeare's play Troilus and Cressida, written around 1600. For Ulysses the hierarchical order inscribed in the cosmos, in which "the heavens themselves, the planets and this centre/Observe degree, priority and place", with the Sun "enthron'd" in "noble eminence" is echoed in human and domestic society. Degree or hierarchy is the key to everything:

---

[1] https://www.ben.edu/degree-programs/LillyConference/upload/Tracey.pdf
[2] http://dhspriory.org/thomas/DeRegno.htm

"How could communities/Degrees in schools and brotherhoods in cities,/ Peaceful commerce from dividable shores,/The primogenitive and due of birth,/Prerogative of age, crowns, sceptres, laurels/But by degree stand in authentic place?" Equally, when hierarchy is disturbed, anarchy breaks loose: "Take but degree away, untune that string,/And hark, what discord follows."[3]

Profoundly conservative in its social implications, the metaphor of the chain of being "served to express the unimaginable plenitude of God's creation, its unfaltering order, and its ultimate unity."[4] The great chain links every atom of nature in an unbroken order. Ulysses warns of the consequences when that order is broken, when "raging of the sea, shaking of earth,/Commotion in the winds, frights, changes, horrors,/Divert and crack, rend and deracinate/The unity and married calm of states/Quite from their fixture."[5]

The New Science and Philosophy: Copernicus, Galileo, Bacon, Descartes

Against this stable hierarchical view, the unsettling investigations and speculations of pioneering modern science and philosophy (there was of course no distinction between the two until the nineteenth century) broke like a series of storms, eventually undermining the foundations.

If priority is given to the work of Nicolaus Copernicus, that is not simply in honour of the major European programme of Earth Observation which has been renamed after him. Copernicus, the son of a well-to-do copper merchant from Torun in Poland, and the nephew of the powerful Bishop of Varmia, was an extraordinary polymath, a scholar, linguist, physician, administrator and currency reformer, whose most far-reaching work was in the field of astronomy and cosmology.

Copernicus was a Renaissance man, who saw the radical potential of a renewed attention to Greek and Latin authors, in the same way as the great Renaissance artists and architects also returned to classical sources. Much of Copernicus's most important work was done at the University of Bologna, where he spent four years supposedly studying canon law but in fact devoting most of his energies to the study of mathematics (under Domenico Maria Novara) and astronomy, using the combined approaches of close attention to the ancient texts (he was one of the first Poles to master ancient Greek) and astronomical observation.

---

[3] Shakespeare, *Troilus and Cressida* I.iii.
[4] E.M.W. Tillyard, *The Elizabethan World Picture* (London: Chatto and Windus, 1943) p. 23.
[5] Shakespeare, op. cit.

On 9 March 1497 he conducted a famous observation of the occultation of the bright star Aldebaran by the moon. Copernicus's mastery of ancient Greek allowed him to read the ancient Greek mathematicians and astronomers in the original language. He started from a position of great respect for the ancient authors, including for their speculative boldness, but also noted the discrepancies between their observations and modern ones. "Although Claudius Ptolemeus the Alexandrian, who in his admirable understanding and exactitude far surpasses the others, with the help of more than four-hundred-years-old observations brought this science almost to its perfection, so that nothing seemed to remain that he had not touched upon; yet we see so many things that do not tally with what should have taken place according to his theory, and this is because certain other motions were discovered later that were unknown to him".[6]

The classical authors also gave Copernicus permission, as it were, to speculate more boldly. In particular he found in the Pythagorean philosopher Philolaus the idea that instead of standing still the earth might be "moved about the element of fire in an oblique circle."[7] He also found in Heraclides of Pontus and Ecphantus the Pythagorean the notion of the earth "turning on its own axis."[8] These earlier speculations allowed him to "meditate on the mobility of the earth."[9]

Some time before 1514 Copernicus wrote a preliminary outline of his heliocentric theory generally known as Nicolai Copernici de hypothesibus motum coelestium a se constitutis commentariolus. This postulated that all the heavenly spheres move around the sun. Copernicus was in no hurry to publish his results, aware that they might be greeted with scorn, disbelief or hostility. He spent his middle years engaged in a variety of worldly pursuits, including administration, practising as a physician, and working to reform the currency, while continuing with his more private studies in his tower at Frauenburg.

From 1539, when he was already an old man, Copernicus was assisted by a brilliant young Protestant professor, Georg Joachim von Lauchen, who took the name Rheticus. Rheticus became Copernicus's disciple and wrote a short summary of the master's theory under the title Narratio Prima or First Account, which was published as a pamphlet and sent to various leading scholars.

---

[6] J.G.Crowther, *Six Great Scientists* (London: Hamilton, 1955), p. 32.
[7] Ibid, p. 34.
[8] Ibid.
[9] Ibid.

Despite this Copernicus did not agree to publish the final version of his great treatise De revolutionibus orbium coelestium until shortly before his death in 1543; tradition has it that he received the final printed pages on the day he died.

Copernicus also took the precaution of dedicating his revolutionary treatise to Pope Paul III, who had a particular interest in astrology. Perhaps because of this, and Copernicus's generally diplomatic cast of character and language, his work initially was quite favourably received by the Catholic Church. Ironically it was much less favourably received by Martin Luther, who referred to it in scathing terms in an after-dinner speech: "This fool wants to turn the whole art of astronomy upside down! But as Holy Scripture testifies, Joshua ordered the sun to stand still, and not the earth".[10]

The full implications of the Copernican revolution took more than half a century to become manifest. This was partly because Copernicus had no great expertise in either calculation or astronomical observation. His treatise did not yield useful practical results, for example in calculating the orbits of the planets, because he assumed that the planets revolved in perfect circles. It was not until Kepler discovered that the planets' orbits were elliptical that more accurate results were obtained. Beyond that, Copernicus framed his treatise as a hypothesis, starting from ancient authority, which might be confirmed by subsequent observation. It was Galileo's confirmation of the Copernican theory using vastly superior astronomical observation by telescope which brought out its revolutionary implications and also provoked violent opposition from the Church.

Galileo

Galileo Galilei, the son of a musician from a distinguished Florentine family, was born in Pisa in 1564. From early in his life he showed traits and aptitudes very different from those of the mild-mannered Copernicus, in addition to a comparable speculative brilliance: he loved argument, gained the nickname "the Wrangler" and was also extremely gifted as a maker of instruments.

Galileo's restless intellect and prodigious energy led him to make numerous important discoveries before he turned his attention to astronomy, stimulated first by the appearance of a new super-nova star in 1604 and then by the invention of the telescope, which he quickly copied. In 1609 he conducted

---

[10] Ibid, p. 36.

perhaps the most remarkable public demonstration and public relations coup in the history of science when he showed off his new telescope to the dignitaries of Venice from the top of the campanile of St Mark's. The senators found that it was possible to observe ships approaching the harbour two hours before they were visible to the naked eye.

Galileo immediately used the new telescope, which he was able to refine and improve, grinding the lenses himself, to make world-changing astronomical observations. One of the most important was his discovery of the moons of Jupiter. These, as he wrote in Sidereus Nuncius, presented "a notable and splendid argument to take away the scruples of those who [can tolerate] the revolution of the planets around the sun in the Copernican system"[11] but who considered the earth's moon in contradiction of that system. He also discovered the phases of Venus which he considered proof the Copernican theory.

Brilliantly gifted as an inventor, scientist and communicator, Galileo was far less astute as a politician, over-estimating the powers of reason and his own powers of persuasion and underestimating the forces of reactionary retrenchment. A long letter he wrote to his pupil Antonio Castelli on December 21, 1613 began the process which would seal his fate: "Who can set bounds to the mind of man? Who dares assert that he already knows all that in this universe is knowable?...I do not think it necessary to believe that the same God who gave us our senses, our speech, our intellect, would have us put aside the use of those."[12]

A copy of this letter which found its way to Rome caught the attention of the Inquisition, which had remained silent on the question of the Copernican theory for seventy years. In March 1616 the Inquisition announced that Copernicus's notion that the sun was the centre of the world was "false and absurd philosophically, and formally heretical". On March 5, 1616 Copernicus's De revolutionibus orbium coelestium was "suspended until corrected."

Galileo managed to evade the direct attention of the Inquisition for another seventeen years, partly through his friendship with Pope Urban VIII, whom he tried to convert to the Copernican system. Privately Urban wrote "that the Holy Church had not condemned the opinion of Copernicus nor was it condemned as heretical, but only as rash, and, moreover, if anyone could demonstrate it to be necessarily true, it would no longer be rash."[13] But the publication of Galileo's great Dialogue Concerning the Two Chief World Systems

[11] http://people.reed.edu/~wieting/mathematics537/sideriusnuncius.pdf
[12] J.G. Crowther, op.cit, p. 73.
[13] Ibid, p. 75.

in February 1632 changed everything. Urban VIII considered that Galileo had tricked him by publishing the work, and then mocked him by putting his opinions into the mouth of a character called Simplicio, named after an ancient philosopher but with a name sounding suspiciously like "simpleton". Urban appointed a secret commission, composed of people hostile to Galileo's opinions, to report on the book, and its sales were suspended. Shortly afterwards Galileo was tried by the Inquisition (in uniquely comfortable circumstances), forced to recant, in one of the saddest episodes in the history both of science and of the Catholic Church, and found "vehemently suspected of heresy." His punishment, relatively mild by the standards of the Inquisition, was to be held under house arrest until his death in 1641.

Galileo's *Dialogue Concerning the Two Chief World Systems* is not just a momentous work in the history of science but a fascinating case-study in how to advocate new or contentious ideas in the face of powerful vested interests. In that sense it continues to have relevance today, in particular for the debate around climate change or global warming.

Francis Bacon: The New Philosophy

The implications of the Copernican revolution and of Galileo's discoveries and confirmations of it were not just narrowly "scientific" in our modern sense. They were paradoxical in both decentring and centring humanity, as far as our place in the universe was concerned. The Copernican theory and its proof showed that the earth was not the centre of the universe, as had been previously been believed by most, if not all, thinkers. It also showed that the earth was not fixed, but moved, and not just in a single but in a double motion, orbiting the sun and spinning on its own axis. And Copernicus correctly surmised that the apparent immobility of the stars demonstrated their great distance from the earth. "What follows from this demonstration," he wrote, "is that the heavens are infinite, in relation to the earth. The extent of this immensity we do not know at all."[14]

In some ways this knowledge might seem to dethrone humanity from its quasi-godlike position, at the centre of the universe and near the head of the hierarchy of being, and also in possession of unquestionable knowledge. But the new discoveries also unleashed hugely liberating and far-reaching energies. They had been made by the unaided intellects of a few master-spirits, with the

---

[14] Ibid, p. 44.

freedom of mind both to re-examine ancient texts and to make new observations, employing the latest technologies.

Not only was the universe infinite; so also, it seemed, were the powers and scope of the human mind. There began, or resumed, a fierce struggle to free human intellect and speculation, exercised in the service of humanity, from the dead hands of custom and established authority. That this would set the master-spirits on a collision course with the reactionary elements of the Catholic Church was shown by the case of Galileo, and before him of Giordano Bruno, burnt at the stake for his heretical beliefs, which included heliocentrism and the idea that the sun was just one of many stars orbited by planets, by order of the Inquisition in 1600.

The philosopher who pursued these implications most energetically was the English statesman, lawyer and author Francis Bacon. Bacon saw that the "advancement of knowledge", as he called it, in the West had been held back by excessive reverence for certain classical texts, especially those of Aristotle, and an over-literal approach to the Bible, combined with a strong bias towards the theoretical—the idea of ivory-tower philosophising—rather than the practical. Bacon proposed a new kind of knowledge based on empirical observation—inductions as opposed to deduction—and dedicated to "easing man's estate".

Even if the earth was no longer the centre of the universe, Bacon wrote that "man, if we look to final causes, may be regarded as the centre of the world; insomuch that if man were taken away from the world, the rest would seem to be all astray, without aim or purpose, to be like a besom without a binding, as the saying is, and to be leading to nothing. For the whole world works together in the service of man; and there is nothing from which he does not derive use and fruit. The revolutions and courses of the stars serve him both for distinction of the seasons and distribution of the quarters of the world. The appearances of the middle sky afford him prognostications of weather. The winds sail his ships and work his mills and engines. Plants and animals of all kinds are made to furnish him either with dwelling and shelter or clothing or food or medicine, or to lighten his labour, or to give him pleasure and comfort; insomuch that all things seem to be going about man's business and not their own."[15]

No clearer or more single-minded exposition of the anthropocentric position could be imagined. But Bacon went still further, beyond the position that everything in nature is constituted in the service of humanity. Bacon proposed that nature should be actively and if necessarily violently hounded (his own word) to reveal her secrets. The eco-feminist writer Carolyn

---

[15] Francis Bacon, *Of the Wisdom of the Ancients*, ch. 26.

Merchant notes the use of female metaphors to describe both nature itself and the subjugation of nature by man, which she compares to the torture of witches. "She [Nature] is put in constraint, molded and as it were made new by art and the hand of man; as in things artificial nature takes orders from man and works under his authority."[16]

The Mechanisation of Nature: Descartes

Bacon's language and use of metaphor enacts a momentous shift: you can see nature changing from a living being, the nurturing Great Mother of antiquity, into a sort of machine. This mechanised view was made more explicit by René Descartes. Descartes applied this idea to the human body: "If the body of man be considered a kind of machine, so made up of bones, nerves, muscles, veins, blood and skin, that although there were in it no mind, it would still exhibit the same motions which it at present manifests involuntarily."[17]

A distinction is still implied here between the human body, capable of being viewed as a machine, and the soul, whose existence Descartes did not doubt. But in the case of non-human nature, there is no soul and therefore no objection or limit to an entirely mechanical view.

Descartes in his great *Discourse on Method* also took further Bacon's championing of a more practical kind of knowledge. He proposed instead "of the Speculative Philosophy usually taught in the Schools, to discover a Practical, by means of which knowing the force and action of fire, water, air, the stars, the heavens and all the other bodies that surround us, as distinctly as we know the various crafts of our artizans, we might also apply them in the same way to all the uses to which they are adapted, and thus render ourselves the lords and possessors of nature."[18]

Dissenting Voices: More, Hermeticism

The Baconian and Cartesian view of nature as either mechanical or almost infinitely pliable, and entirely subject to man, which would become the dominant one for centuries to come, up to our own age, was not universally accepted.

---

[16] See Carolyn Merchant, *The Death of Nature: Women, Ecology and the Scientific Revolution* (New York: Harper & Row, 1989).

[17] René Descartes, *Meditation VI*.

[18] René Descartes, *Discourse on Method*.

One of the interesting dissenting voices is that of the Cambridge Platonist philosopher Henry More, who conducted a correspondence with Descartes in the 1640s. In one letter More writes to Descartes: "I recognize in you a subtle keenness, but also, as it were, the sharp and cruel blade which in one blow, so to speak, dared to despoil of life and sense practically the whole race of animals, metamorphosing them into marble statues and machines."[19]

More is speaking here for the whole older tradition which held, with Plato's *Timaeus*, that the world was not dead matter but a living being. Part of that tradition is the Hermetic one.

Hermeticism was an alternative mystical philosophy based on the writings of Hermes Trismegistus which regained considerable influence during the Renaissance. The Hermetic texts, which were thought to be of very ancient origin, possibly the work of an Egyptian priest, but in fact probably date from 200 to 500 A.D., were translated by the Renaissance humanists Ficino and Pico della Mirandola and informed the thought of Giordano Bruno. Hermeticism posits both the unity of all things and the all-importance of the invisible world. A part of Hermeticism is alchemy, seen not just as the physical transformation of lead into gold but as an enquiry into the spiritual composition of all aspects of nature.

Newton

The great summation of the seventeenth century scientific revolution is found in Sir Isaac Newton's *Principia Mathematica*, bringing together the astronomical discoveries of Copernicus, Kepler and Galileo, Galileo's pioneering work on the laws of motion and Newton's own discoveries regarding gravity.

Newton's vision of a kind of clockwork universe, operating according to regular and discoverable principles, had a huge influence on his contemporaries (the poet Alexander Pope wrote "Nature and Nature's Laws lay hid in Night/God said, "Let Newton be!" and all was light"[20]) and dominated thinking in the West for over 200 years. What had previously been seen as heretical became the new orthodoxy; now it was the turn of a few mystics and Romantic poets to challenge the new orthodoxy.

---

[19] Henry More, *Letters to Descartes*.
[20] Alexander Pope, *Epitaph on Sir Isaac Newton*.

Kepler's Harmony of the World and Newton's Alchemical Works

For all the advances in understanding of the new scientific age, we should remember that the world view of these scientific pioneers was by no means strictly scientific, in the sense we have become familiar with. For example, Johannes Kepler's *Harmony of the World (Harmonices Mundi)* is an extraordinary attempt to make a link between musical harmony and the dispositions and orbits of the planets, or to make real astronomical sense of what since ancient times had been called "the music of the spheres". Kepler found that the difference between the minimum and maximum angular speeds of a planet in its orbit could be closely related to musical intervals. In the case of the Earth, the difference in orbiting speeds approximated to the interval of a semitone, and Kepler commented with doleful irony: "The Earth sings Mi, Fa, Mi: you may infer even from the syllables that in this our home misery and famine hold sway."[21] Kepler also held that the Earth has a soul because it is subject to astronomical harmony. "I may say with truth that whenever I consider in my thoughts the beautiful order, how one thing issues out of and is derived from another, then it is as though I had read a divine text, written into the world itself, not with letters but rather with essential objects, saying: Man, stretch thy reason hither, so that thou mayest comprehend these things."[22] Even Isaac Newton, two generations after Kepler, devoted much of his time to abstruse alchemical researches. He has even been called "the last of the magicians."

Globes

Also straddling the divide between what we now call science and art are the beautiful globes, both terrestrial and celestial, produced in Europe and the Islamic world from the time of the Renaissance onwards. Some of the most magnificent were made in Venice in the late seventeenth and early eighteenth centuries by Vincenzo Coronelli. It is hard not to agree with the connoisseur Peter Allmayer-Beck, some whose collection of globes is displayed, on loan, in the Globe Museum in Vienna, when he writes: "Contemplating the highly imaginative mythological images of constellations on old celestial globes, I usually feel slightly nostalgic, representation on post-1850 globes reduces the heavens to a technically abstract universe."[23]

---

[21] Johannes Kepler, *Harmonices Mundi.*
[22] Johannes Kepler, *Calendar for the Year 1604.*
[23] Globe Museum, Vienna.

4

The Enlightenment, The Romantic Rebellion, The Industrial Age, The Nature Conservation Movement, The Twentieth Century and Total War

The period of the Enlightenment was dominated by Newton's mechanistic view of Nature. As we have seen, the poet Alexander Pope spoke of Newton's discoveries, perhaps with a certain tinge of irony, in quasi-divine terms, as comparable to the original Creation. But despite the general sense that Newton's physics had convincingly explained the workings of the universe as a sort of vast machine following universal laws, some of the key thinkers of the Enlightenment put forward alternative ideas.

Diderot

One of the most striking comes in Denis Diderot's philosophical dialogue *D'Alembert's Dream*. At the start of the dialogue, Diderot argues for a kind of latent sensitivity in all matter ("stone must feel"), in terms which prefigure the discoveries of modern physics. Diderot presents himself as a materialist, who rejects conventional distinctions between body and soul—a distinction which as we remember was still upheld by both Descartes and Newton. This thoroughgoing materialism could lead in two directions: either to the notion that everything is raw material, with no spiritual dimension, or to the idea put forward at the beginning of Diderot's conversation with D'Alembert, that all matter feels and has sensitivity. These two directions clearly have hugely

© Springer International Publishing Switzerland 2017
H. Eyres, *Seeing Our Planet Whole: A Cultural and Ethical View of Earth Observation*, DOI 10.1007/978-3-319-40603-9_4

important implications in terms of our present study: the former leading to the idea that everything, including human beings, is what Heidegger would term "standing reserve", awaiting transformation and exploitation, the latter returning to the Platonic idea of the earth as a living being.

Swift

Another maverick Enlightenment figure is the Anglo-Irish satirist Jonathan Swift. In Gulliver's Travels, the hero, Lemuel Gulliver, discovers and is rescued by the flying island of Laputa, a realm devoted to the pursuit of madcap scientific schemes. These include trying to extract sunbeams from cucumbers and softening marble for use in pillows. Laputa is often seen as a satire on experiments pursued by the Royal Society.

Swift's target is the kind of scientific experimentation which has lost touch with practicality. Although one might criticise Swift for failing to anticipate the extraordinary scientific breakthroughs which have "eased man's estate", his satire is still acute in relation to certain contemporary projects which are avidly supported or pursued in defiance of the kind of common sense which recognises limits: for instance, the much trumpeted ideas of colonising Mars, or some of the more extreme forms of geo-engineering.

The particular schemes mocked by Swift may now seem merely quaint, but his psychological insight into the mindset of a certain type of scientist or mathematician remains remarkably prescient. The inhabitants of Laputa go around "under continual Disquietudes, never enjoying a Minute's Peace of Mind; and their Disturbances proceed from Causes which very little affect the rest of Mortals. Their Apprehensions arise from several Changes they dread in the Celestial Bodies. For Instance: that the Earth by the continual Approaches of the Sun towards it, must in course of Time be absorbed or swallowed up."[1] Their fear of these "impending Dangers" means that the Laputians can "neither sleep quietly in their Beds, nor have any relish for the common Pleasures or Amusements of Life."[2]

The Laputians have their "minds on other things" as the common phrase has it. Being perpetually distracted and having one's mind on other things

---

[1] Jonathan Swift, *Gulliver's Travels*, Part 3, Ch. 2 https://www.gutenberg.org/files/829/829-h/829-h.htm
[2] Ibid.

may be a characteristic of the kind of person who is unable to dwell comfortably in one place; by extension, of the mindset which is dissatisfied with the confines of earth and bent on colonising other planets.

The Romantic Rebellion

The Romantic movement, led by poets and philosophers, starting in the last two decades of the eighteenth century, can be seen as a passionate rebellion against the mechanistic view of nature associated with Descartes and Newton. But the movement was complex and contained many disparate threads.

Wordsworth

The English poet William Wordsworth (1770–1850), who absorbed contemporary currents in German philosophy through his close friend Samuel Taylor Coleridge, states the case in its purest form in his poem 'The Tables Turned':

> Sweet is the lore which Nature brings;
> Our meddling intellect
> Mis-shapes the beauteous form of things:-
> We murder to dissect.
>
> Enough of Science and of Art;
> Close up those barren leaves;
> Come forth, and bring with you a heart
> That watches and receives.[3]

The strongest word in these lines is "murder": Wordsworth sees the detached scientific curiosity which takes things apart in order to understand them better as a kind of deadly violence. The torture inflicted on nature, implicit in Bacon, has had fatal results. It is also worth noting the last two lines of the poem, in which Wordsworth calls for a different approach, based on the living appreciation of nature, outdoors rather than in the study or the laboratory ("Come forth") and on the sensitive perceptions and intuitions of the heart rather than the head.

---

[3] William Wordsworth, *Poetical Works* (London: Oxford University Press, 1904), p. 481.

Wordsworth's poetry is one long hymn to the power and beauty of nature, not as analysed and dissected by science and mathematics but as experienced in the living mind and soul. Wordsworth rejects the dualism of Descartes, which relegates nature to the status of inert matter, and the mechanical view which reached its apogee in Newton. Instead, he feels "a sense sublime/Of something far more deeply interfused,/Whose dwelling is the light of setting suns,/And the round ocean and the living air,/And the blue sky, and in the mind of man."[4] Nature and man are not separate but infuse and interpenetrate each other, though this is something not amenable to scientific proof, but an intuition which must be "felt."

Wordsworth grew up in the English Lake District, an area of outstanding natural beauty, and his earliest memories are recounted in the great long autobiographical poem *The Prelude*, subtitled The Growth of a Poet's Mind. In a famous episode, Wordsworth tells how as a young boy, on a summer's evening, he went rowing on "the shores of Patterdale". Suddenly, "a huge cliff,/As if with voluntary power instinct,/Upreared its head. I struck and struck again,/And growing still in stature the huge cliff/Rose up between me and the stars, and still/With measured motion, like a living thing/Strode after me."

What is striking is that Wordsworth does not dismiss this experience as childish fancy, but takes his youthful sense of nature as mysteriously alive and even full of purpose entirely seriously. The mountains and the lakes meant as much to Wordsworth—perhaps even more—than the presence of other human beings; this "deep green" attitude would influence such later thinkers as Henry David Thoreau and John Muir, the founder of the National Park movement in the USA. Its dark, shadow side is the attitude which places more value on nature and landscape than on human beings.

Goethe

Wordsworth's reverence towards the mystery, power and animate force of nature was shared by his German contemporary, the polymath Johann Wolfgang von Goethe. But Goethe was not just a poet, but also a pioneering

---

[4] William Wordsworth, 'Tintern Abbey'.

and unconventional scientist. Rather than rejecting science as Wordsworth seemed to in his more intemperate moments (though not in the 1802 Preface to Lyrical Ballads), Goethe envisaged and practised a new kind of holistic science, very different from the mechanistic variety associated with Newton. He made serious contributions in the fields of anatomy, botany, geology, optics, colour theory and meteorology. Perhaps the most famous and controversial is his *Theory of Colours* (*Farbenlehre*), in which he attempted to disprove Newton's theory as expounded in the *Opticks* (1704). Only recently has Goethe's theory attracted serious attention, not as a refutation of Newtonian optics, but as starting from a different standpoint, attempting to understand colour as perceived by the human eye rather than as an objective phenomenon in the world.

Goethe was also a pioneering meteorologist, much influenced by the Englishman Luke Howard's An Essay on the Modification of Clouds. His studies acquired a more practical orientation when Duke Carl August of Weimar, whom Goethe had earlier served as Prime Minister, decided in the 1810's to establish a series of weather stations scattered around his duchy.

Far from being a detached observer, motivated purely by curiosity, the scientist as envisaged by Goethe is very much part of what he or she sees, observes, experiences. Rather than putting the stress on instruments and techniques, Goethe returns science to its point of origin, the fully alive and sensitive human being. As he wrote in his 1792 essay "The Experiment as Mediator between Subject and Object", Goethe believed that "the human being himself, to the extent that he makes sound use of his senses, is the most exact physical apparatus that can exist."

The Modern Prometheus

The Romantic movement of the late eighteenth and early nineteenth centuries, as we observed, contains many strands. One of them is a renewed interest in the figure of Prometheus. Such luminaries as the German composer Ludwig van Beethoven and the English poet Percy Bysshe Shelley (in his play *Prometheus Unbound*) were drawn to Prometheus as a symbol of the human quest and fight for freedom against the tyranny of political and religious autocracy.

But perhaps the most interesting and prescient treatment of the Prometheus myth in the Romantic period comes in the novel written by Shelley's wife, Mary Shelley, and entitled *Frankenstein, or The Modern Prometheus*.

Mary Shelley's Gothic tale concerns a passionate and promising scientist, Victor Frankenstein, who succeeds in creating a monster in human form. Frankenstein is disgusted by his creation and attempts to escape from it. However, the monster, rejected by people he tries to befriend because of his hideous appearance, pursues his creator, and pleads with him to create a female companion. Frankenstein does so, but kills the female, fearing a new breed of monsters might emerge. The monster wreaks a terrible revenge, murdering Frankenstein's newly wedded bride, Elizabeth.

Mary Shelley's intriguing tale is open to a wide variety of interpretations. One thing it is not, however, is an optimistic paean to the powers of science. The tale certainly raises questions about the responsibility of science, and the unforeseen and frightening implications of scientific breakthroughs.

The Industrial Revolutions and The Industrial Age

The Romantic movement in literature, philosophy, art and music coincided with the beginnings of what has turned out to be the most dynamic period in human history. The innovations of the industrial revolution in textile manufacturing, iron production, various chemical processes, the use of water power, the change from wood to coal as fuel, had and continue to have the most far-reaching effects. Among these effects were unprecedented increases in world production of goods and services, extractive industries and, eventually, population. The industrial revolution ushered in the modern period of economic growth and increased living standards for the masses. The increase in population, especially from the 1870s, is startling. Worry about this was voiced as early as 1798 by the Rev. Thomas Malthus in *An Essay on the Principle of Population*. Malthus argued first that "in all societies...there is a constant effort towards an increase in population", and then that because population increases geometrically and food production arithmetically, this increase is bound to lead to food shortages and starvation. The pessimistic Malthus has been criticised for failing to factor in the effects of technological innovation, which so far have enabled world food production to keep pace with population increase. However, his broader point about the growing pressure on resources in a finite world still seems to be relevant.

The industrial revolution also ushered in the age of socialism; without it the Communist revolutions in Russia and China are unthinkable.

The industrial revolution started, by common consent, in the manufacturing towns of the north of England. Here the new mechanised processes of textile spinning and weaving had their most dramatic effects.

The Luddites

The immediate impacts of these innovations were not necessarily seen as positive. The Luddites were a more organised version of groups of disgruntled textile workers who began smashing industrial machinery in the late eighteenth century, giving rise to legislation such as the Protection of Stocking Frames etc. Act of 1788.

More narrowly defined, "the Luddites were the people who broke machinery as deliberate, calculated policy in a particular historical period, the years 1811–16."[5] The Luddite "revolt" was certainly quite dramatic, as were the means used to suppress it: "the army of 12,000 required for domestic use to suppress the Luddites was a greater force than Wellington had taken to Portugal in 1808."[6]

Recent historians have stressed that the Luddites were not opposed to machinery per se, and certainly not to the stocking-frame, which had been in use since Elizabethan times; they wanted to preserve their jobs and crafts at a time of terrifying unemployment and starvation, long before the creation of any kind of "welfare state". Thus the historian Eric Hobsbawm described their activities as "collective bargaining by riot".[7] Perhaps even more impressively, the Luddites were defended by Lord Byron in his maiden speech to the House of Lords in 1812:

"Whilst these outrages must be admitted to exist to an alarming extent, it cannot be denied that they have arisen from circumstances of the most unparalelled distress. The perseverance of these miserable men in their proceedings, tends to prove that nothing but absolute want could have driven a large and once honest and industrious body of the people into the commission of excesses so hazardous to themselves, their families, and the community. At the time to which I allude, the town and county were burdened with large detachments of the military; the police was in motion, the magistrates assembled, yet all these movements, civil and military had led to—nothing. Not a single instance had occurred of the apprehension of any real delinquent actually taken in the fact, against whom there existed legal evidence sufficient for conviction. But the police, however useless, were by no means idle: several notorious delinquents had been detected; men liable to conviction,

---

[5] Malcolm I. Thomis, *The Luddites: Machine-Breaking in Regency England* (Newton Abbot: David & Charles, 1970), p. 27.
[6] Ibid, p. 144.
[7] Quoted in ibid, p. 35.

on the clearest evidence, of the capital crime of poverty; men, who had been nefariously guilty of lawfully begetting several children, whom, thanks to the times!—they were unable to maintain. Considerable injury has been done to the proprietors of the improved frames. These machines were to them an advantage, inasmuch as they superseded the necessity of employing a number of workmen, who were left in consequence to starve. By the adoption of one species of frame in particular, one man performed the work of many, and the superfluous labourers were thrown out of employment. Yet it is to be observed, that the work thus executed was inferior in quality, not market-able at home, and merely hurried over with a view to exportation. It was called, in the cant of the trade, by the name of Spider-work. The rejected workmen, in the blindness of their ignorance, instead of rejoicing at these improvements in arts so beneficial to mankind, conceived themselves to be sacrificed to improvements in mechanism. In the foolishness of their hearts, they imagined that the maintenance and well doing of the industrious poor, were objects of greater consequence than the enrichment of a few individuals by any improvement in the implements of trade which threw the workmen out of employment, and rendered the labourer unworthy of his hire. And, it must be confessed, that although the adoption of the enlarged machinery, in that state of our commerce which the country once boasted, might have been beneficial to the master without being detrimental to the servant; yet, in the present situation of our manufactures, rotting in warehouses without a pros-pect of exportation, with the demand for work and workmen equally dimin-ished, frames of this construction tend materially to aggravate the distresses and discontents of the disappointed sufferers. But the real cause of these dis-tresses, and consequent disturbances, lies deeper. When we are told that these men are leagued together, not only for the destruction of their own comfort, but of their very means of subsistence, can we forget that it is the bitter policy, the destructive warfare, of the last eighteen years, which has destroyed their comfort, your comfort, all men's comfort;—that policy which, originating with "great statesmen now no more," has survived the dead to become a curse on the living unto the third and fourth generation! These men never destroyed their looms till they were become useless, worse than useless; till they were become actual impediments to their exertions in obtaining their daily bread."[8]

If Byron's speech stands out for its magnificent irony and moral power, Karl Marx's comment on the Luddites in *Das Kapital* Vol. 1 is perhaps more

---

[8] http://www.luddites200.org.uk/LordByronspeech.html

prescient: he criticised the Luddites for attacking "the material instruments of production" rather than "the form of society which utilizes those instruments."

Another impact was a drastic and unprecedented increase in pollution. No better description of this exists than Charles Dickens' description of the place he calls Coketown (often identified with Manchester) in his novel *Hard Times*: "it was a town of red brick, or of brick that would have been red if the smoke and ashes had allowed it; but as matters stood it was a town of unnatural red and black like the painted face of a savage. It was a town of machinery and tall chimneys out of which interminable serpents of smoke trailed themselves for ever and ever; and never got uncoiled. It had a black canal in it, and a river that ran purple with ill-smelling dye, and vast piles of building full of windows where there was a rattling and a trembling all day long, and where the piston of the steam-engine worked monotonously up and down, like the head of an elephant in a state of melancholy madness. It contained several large streets very like one another, and many small streets still more like one another, inhabited by people equally like one another, who all went in and out at the same hours, with the same sound on the same pavements, to do the same work, and to whom every day was the same as yesterday, and to-morrow, and every year the counterpart of the last and the next."

Dickens' description of Coketown is memorable for many reasons, but one of them is that the kind of townscape, landscape and society he is portraying, now more or less familiar, appears as a sort of novel monstrosity. Nothing quite like this had ever been seen on the surface of the earth; no houses so smoke-blackened, no stretch of water as thoroughly and unnaturally poisoned and polluted, no society of people so rigidly regimented.

The industrial revolution had, and continues to have, by far the greatest impact on the earth, the oceans and the atmosphere, of any set of events in human history. According to some scientists, it has ushered in a new era in earth's history defined by the human influence on the earth, the Anthropocene Era.

The long-term impacts of the industrial revolution are just as dramatic as the short-term ones noted by Dickens. By facilitating the unprecedented increases in the world's productive output and population, the industrial revolution also created the conditions for the increased flow of greenhouse gases into the atmosphere. The rises in concentrations of CO_2, methane and other gases measured by scientists using ice core samples and other means begin around the time of the industrial revolution in the late eighteenth century. From that point they have shown a steady increase.[9]

---

[9] See www.acs.org

The Industrial Age and The Victorian Reform Movement: Ruskin, William Morris

The polluting effects of the industrial revolution and its transformation of human beings into (as many saw it) the slaves of machines generated a passionate counter-movement, led by such luminaries as Dickens, Ruskin and William Morris. By extension, it provoked the beginnings of environmental regulation and policy, and the first steps towards the conservation of nature. At first this was aimed mainly at cleaning up the most drastic effects of industrialisation, in terms of water, air and terrestrial pollution. In none of these areas was progress especially speedy. In the UK the Clean Air Act was only passed as recently as 1956. An earlier landmark was the construction of the London sewers under the direction of Joseph Bazalgette following the "great stink" of 1858.

John Ruskin and William Morris focussed much of their attention on the dehumanising effects of industrial modes of production. Ruskin was also a pioneering environmentalist. He revered Wordsworth, built a house on the shores of Coniston Water in the English Lake District and painted many exquisitely detailed watercolour studies of natural phenomena. He was also one of the first men of the nineteenth century to perceive and anticipate the threats to the environment posed by the industrial revolution. He has been called "the first ecologist".

Like Ruskin William Morris was profoundly interested in the forms and beauties of nature (for instance in his floral textile designs) and passionately critical of the effects of industrial mass production both on human beings and on the natural environment. He wrote prophetically about science teaching "Manchester how to consume its own smoke, or Leeds how to get rid of its superfluous black dye without turning it into the river."[10] He spoke also of "the latest danger which civilisation is threatened with, a danger of her own breeding: that men in struggling towards the complete attainment of all the luxuries of life for the strongest portion of their races should deprive their whole race of all the beauty of life."[11]

William Morris was a socialist, influenced by the writings of the most important socialist thinker, Karl Marx. Marx's main focus was on the human and social effects of capitalism, but he also showed a prescient awareness of the actual and potential effects of capitalist exploitation on the natural world,

[10] William Morris, *Hopes and Fears for Art.*
[11] Ibid.

as in the following statement from *Das Kapital*: "Even an entire society, a nation, or all simultaneously existing societies taken together, are not owners of the globe. They are simply its possessors, its usufructuaries, and, like boni patres familias, they must hand it down to succeeding generations in an improved state in an improved condition."[12] Marx also expressed scepticism about the long-term sustainability of industrial farming methods: "all progress in increasing the fertility of the soil for a given time is a progress towards ruining the more long-lasting sources of that fertility. The more a country starts its development on the foundations of modern industry, like the United States, for example, the more rapid is the process of destruction. Capitalist production, therefore, develops technology…only by sapping the original sources of all wealth—the soil and the worker."[13]

Friedrich Engels also evinces a concern about the effects on nature, not just on industrial workers, of industrial modes of production.

"Let us not flatter ourselves overmuch on account of our human victories over nature. For each victory nature takes its revenge on us. Each victory, it is true, in the first place brings about the results we expected, but in the second and third places it has quite different, unforeseen effects which only too often cancel out the first."[14]

The Nature Conservation Movement

At a less theoretical level, the Victorian era in Britain and the latter part of the nineteenth century more widely also brought forth the first concerted movements in nature conservation, as a response to the untrammelled exploitation and despoliation of the natural environment. In the United States, the indefatigable, Scottish-born naturalist and campaigner John Muir spearheaded the National Park movement, to protect certain especially valuable ecosystems, which led to the creation of Yosemite National Park in 1890 (followed by Sequoia, Grand Canyon and Mount Rainier National Parks) and the US's leading nature conservation NGO, the Sierra Club, in 1892. In The Yosemite, published in 1912, Muir wrote that "no temple can compare with Yosemite. Every rock in its wall seems to glow with life." Here we have an echo of the view of the earth as living being familiar from Plato and other early thinkers.

---

[12] https://www.marxists.org/archive/marx/works/1894-c3/ch46.htm
[13] https://www.marxists.org/archive/marx/works/download/pdf/Capital-Volume-I.pdf
[14] Friedrich Engels, *The Origins of the Family, Private Property and the State.*

In England, Wales and Northern Ireland the National Trust for Places of Historic Interest or Natural Beauty was founded in 1895 by Octavia Hill, Sir Robert Hunter and Hardwicke Rawnsley. Here once again the impulse was to protect and preserve, for the enjoyment, recreation and spiritual restoration of everybody, as many as possible of the country's most valuable and beautiful (and threatened) landscapes.

The Twentieth Century: World War, Total Devastation, the Rise of the Environmental Movement

The end of the nineteenth century could be seen as in some ways returning to its beginning. Humanity had unleashed new and astonishing powers, with the potential to transform lives (often for the better) and nature (usually for the worse). Material wealth had increased enormously and living conditions for the majority had almost certainly also improved (though some scholars point out that wages only began to rise for the majority of British workers in the 1870s, while at the same time a long agricultural slump had begun). The worst excesses of unfettered capitalism, both at the level of human society, with the rise of the Trade Union movement and legislation to limit working hours and provide holiday time, and of the natural world, with the rise of the conservation movement, were being at least addressed.

But the twentieth century, after the deceptively rosy dawn of the Edwardian period, would bring the unprecedented destruction of the two world wars. The historian Eric Hobsbawm, in The Age of Extremes, describes the twentieth century as "without doubt the most murderous century of which we have record". World War One was initially seen as a limited conflict which would only last months. It became a global cataclysm lasting more than four years in which 17 million died.

World War One also offered a vision of a kind of environmental hell. Artists such as Paul Nash depicted a landscape which was not so much devastated as wholly denatured. Entire stretches along the Western Front were reduced to fetid mud, littered with unburied corpses, where all life had been obliterated. As in the case of Dickens' Coketown sixty years earlier, nothing like this had been seen in human history. Here was a vision of what human industrial power, turned against itself, could make of the world—a wasteland, which is also the title of T.S. Eliot's poem of 1919, which set the tone for a generation.

At the same time, injured soldiers were returning home, with wounds so ghastly—as recorded by such painters as Otto Dix and Georg Gross—that they could have come out of the imagination of Hieronymous Bosch. Through a series of diplomatic muddles and stalemates, humanity had succeeded only in creating hell on earth.

In Russia the end of World War One coincided with the Bolshevik Revolution (after an earlier inconclusive attempt at revolution in 1905), one of the most significant events in history. The Russian Revolution, and the subsequent establishment of a totalitarian regime under Stalin, was seen by many throughout the world as a beacon of hope, the coming of a fairer socialist society which would do away with many of the evils of unfettered capitalism.

The hope was quite quickly shattered, especially by the show trials of the 1930s. And if the hope had included the promise of a more sustainable relationship between human beings and the natural world, that promise would also turn out to be completely without foundation. Despite the difference of economic system—socialism or communism as opposed to capitalism—the relationship of human beings to the natural world, as merciless exploiters and subjugators, remained unchanged. Some of the worst examples of environmental despoliation on record, such as the virtual disappearance of the Aral Sea and the Chernobyl nuclear disaster, happened under the watch of Soviet communism.

In other parts of Europe, and the Western World, the 1920s and 1930s were dominated by, on the one hand, a feverishly hedonistic attempt to forget the horrors of war (the Roaring Twenties), and, on the other, a worldwide depression, triggered by the Wall Street crash of 1929, which, as many recognised very early, would create the conditions for another conflict. The victory of Communism in Russia was answered by the rise of Fascism in Italy and Germany. These ideologies clashed first in the Spanish Civil War of 1936–9.

5

The Post-War Period and the Rise of Ecological Consciousness

The recovery after the Second World War initially and understandably focussed on repairing the damage to cities, towns and infrastructure, on providing shelter and food for tens of millions left destitute. But quite quickly there developed a growing awareness of the intensifying damage being done to the natural world after a hundred years of industrialisation, including the loss of habitats and species. The evolutionary biologist, humanist and first director of UNESCO Julian Huxley was one of the earliest and most prominent public figures to sound the alarm. Huxley was one of the founders of the World Wildlife Fund, later renamed Worldwide Fund for Nature (WWF), established in 1961 in order "to stop the degradation of the planet's natural environment and to build a future in which humans live in harmony with nature". Huxley's earlier experience had included helping to set up nature reserves and national parks in east Africa. This was an extension on a grander scale of initiatives such as the Wildlife Trusts in the UK, which owed much to the pioneering efforts of the Rothschild family. Julian Huxley was also, and more controversially, an advocate of eugenics and of population control. In his essay The Crowded World he predicted, with remarkable accuracy, that the world's population would reach 6 billion by 2000.

© Springer International Publishing Switzerland 2017
H. Eyres, *Seeing Our Planet Whole: A Cultural and Ethical View of Earth Observation*, DOI 10.1007/978-3-319-40603-9_5

Silent Spring and Earthrise: The "First Wave" of Environmentalism

The year after the founding of WWF, the biologist Rachel Carson published *Silent Spring*, possibly the most influential environmental book of all time. Carson warned about the overuse of poisonous pesticides and the build-up of chemical pesticide residues, especially from DDT, going up the food chain through insects, insect-eating birds and eventually raptors. She also accused the chemical industry of spreading disinformation, and public officials in turn of being too uncritical and unquestioning of the chemical industry. The image contained in the title of her book, inspired by the line from Keats's poem La Belle Dame Sans Merci "and no birds sing" of a world denuded of the beauty of birdsong, proved to have a remarkable galvanising power.

Carson herself, who was in declining health when Silent Spring came out in October 1962, never called for an outright ban on DDT, but the pesticide was subsequently prohibited for agricultural use in the USA. She and her publishers had to withstand fierce counter-attacks from the chemical industry, especially DuPont, the manufacturer of DDT, but Carson proved a calm and persuasive communicator. The influence of Silent Spring, which sold more than 500,000 copies, proved far-reaching, not just in the DDT ban but in the wider move towards the establishment of the US Environmental Protection Agency and, beyond that, in the forming for the first time in history of a nationwide civic environmental movement.

Another key moment in the development of the first wave of environmentalism came in 1968 when as part of the Apollo 8 space mission the astronauts William Anders, Frank Borman and Jim Lovell took the famous "Earthrise" photograph of the earth seen from lunar orbit. The emotive and galvanising power of Earthrise is comparable to that of Silent Spring. It has been called by the nature photographer Galen Rowell "the most influential environmental photograph ever taken."

What both Silent Spring and Earthrise conveyed was the sense of the beauty and fragility of the earth, threatened by unrestrained human development. The Earthrise photograph thrillingly portrayed the shining perfection of the earth's form and the rich aliveness implied by its colourfulness, set against the barren rock of the Moon, above which the bright blue earth is seen to rise. Never before had human beings been able to see the earth as a complete orb. In some ways this vision confirmed the intuitions of philosophers such as Plato.

Silent Spring, on the other hand, drew attention to the terrifying unintended consequences of "technological fixes", and, equally important, to the intricate interweaving and interdependence of all aspects of eco-systems, so that a chemical intended to eradicate insect pests could end up wiping out bird species.

The environmental movement, both in the United States and more globally, gathered pace in the late 1960s and early 1970s. This period saw the founding of the two best-known campaigning environmental NGOs, Friends of the Earth and Greenpeace, the former created initially as an anti-nuclear organisation and then as a network of linked groups working upwards from grassroots activism to national pressure-group level and the latter concentrating on high-profile adventurous protests-cum-escapades, such as the anti-whaling campaigns featuring activists in speedboats, designed to capture newspaper headlines and lead stories in new bulletins.

Both eventually evolved into large organisations with millions of members. Another seminal event was the first Earth Day in 1970, the brainchild of the Wisconsin senator Gaylord Nelson. On 22 April of that year, 20 million Americans took to the streets in a nationwide demonstration against the pollution of the environment, in favour of clean air, water, and protection for wildlife. The demonstration was remarkable in that it brought together people of different political persuasion, social strata, race and gender. Earth Day 1970, together with Silent Spring and the Earthrise photograph, is credited with a key role in forging the emerging environmental awareness and ecological consciousness, which led to the forming of the US Environmental Agency (which began to operate in December 1970) and the passing or strengthening of the US Clean Air Act, Clean Water Act (1977) and Endangered Species Act (1973).

The environmental movement was simultaneously gaining ground globally. The first United Nations Conference on the Environment, or Earth Summit, took place in Stockholm in 1972, gathering representatives from 113 countries, 19 inter-governmental agencies and more than 400 inter-governmental and non-governmental organisations. It resulted in the Stockholm Declaration, including among its 26 principles the following:

- Natural resources must be safeguarded.
- The Earth's capacity to produce renewable resources must be maintained.
- Wildlife must be safeguarded.
- Non-renewable resources must be shared and not exhausted.
- Pollution must not exceed the environment's capacity to clean itself.
- Damaging oceanic pollution must be prevented.

This was perhaps the first of many declarations issued by international environmental conferences demonstrating good intentions rather than any firm will to action.

1972 was also the year of the publication of the celebrated and controversial Limits to Growth report by the environmental group The Club of Rome, predicting "overshoot and collapse" of the global system by the mid- to late-twenty-first century unless radical steps were taken to reduce pressure on the environment. Limits to Growth has been criticised, notably by Robert Solow, as an essentially Malthusian analysis, failing to factor in the exponential potential of technology. However, some of the essential findings of Limits to Growth have been confirmed by a number of recent studies, including one carried out in 2014 at the University of Melbourne.[1]

In Europe, Canada, New Zealand and Australia the environmental movement took more directly political form in the founding of Green parties. Europe's first green party, the UK Ecology Party, was founded in 1973 but failed to make much of an impact. The German Green Party, die Gruenen, was founded in the late 1970s and has enjoyed considerable success and influence, despite the well-publicised split between different factions. Die Gruenen contested their first national election in 1980 and won 27 seats in the federal elections in 1983. At various times Die Gruenen have held the balance of power in German coalition governments.

Philosophical Background to Green Politics: Deep Ecology, Arne Naess, Aldo Leopold, Heidegger/ Enframing

Among the philosophical underpinnings of the emerging environmental movement and of European Green Politics are the perspective known as Deep Ecology, incorporating the thinking of the Norwegian philosopher Arne Naess and the American nature writer Aldo Leopold, and the work of the controversial German philosopher Martin Heidegger.

Deep Ecology argues for the intrinsic worth of nature as a whole, and against the idea of natural resources or natural capital envisaged as being purely for the benefit of human beings. The first two principles of the Deep Ecology Platform, devised by Naess and George Sessions, are as follows:

1. The well-being and flourishing of human and nonhuman life on Earth have value in themselves (synonyms: inherent worth, intrinsic value, inherent value). These values are independent of the usefulness of the nonhuman world for human purposes.
2. Richness and diversity of life forms contribute to the realization of these values and are also values in themselves.[2]

---

[1] http://www.theguardian.com/commentisfree/2014/sep/02/limits-to-growth-was-right-new-research-shows-were-nearing-collapse

[2] http://www.deepecology.org/platform.htm

Aldo Leopold for his part proclaimed a "land ethic [which] simply enlarges the boundaries of the community to include soils, waters, plants and animals, or collectively: the land…A land ethic changes the role of *Homo sapiens* from conqueror of the land-community to plain member and citizen of it. It implies respect for his fellow members, and also respect for the community as such."[3]

Perhaps the most important philosophical underpinning of the emerging environmental movement and of European Green politics is the work of the controversial German philosopher Martin Heidegger. Heidegger's early masterwork Being and Time addressed such questions as the essentially temporal nature of existence and the possibility of authentic being-in-the-world. In the 1930s Heidegger was an enthusiastic supporter of the Nazi party, and in 1933 joined the party and delivered an infamous address as Rector of Freiburg University in which he called for the German people to fulfil its historical mission. Heidegger resigned his rectorship the following year and took no further part in Nazi meetings, though he remained a member of the Nazi party until its dissolution in 1945. He also never publicly apologised for or retracted his support for Nazism.

It is Heidegger's later work which has had more influence on environmental thinking. In the essay 'The Question Concerning Technology', Heidegger develops the concept of "enframing" (Gestell). The essence of technology is not technology itself, but a view by which everything is a means to an end, under the sign of "utter availability and sheer manipulability", in the words of David Farrell Krell. Thus the River Rhine becomes no longer a river in the way that a poet such as Hölderlin saw it, but a means of transport or source of hydroelectric energy.

The power of Heidegger's insights lie in his realisation that the problem goes beyond particular manifestations of technology into the fundamentals of our prevailing way of thinking. To take an example, one of the currently touted ways of dealing with the build-up of greenhouse gases in the atmosphere is through geo-engineering. But geo-engineering "solutions" such as pumping aerosols into the upper atmosphere or reflecting back light from the sun risk disrupting not-very-well understood planetary systems of weather and rainfall patterns(for instance, the monsoon rains in India, Pakistan and Sri Lanka) with potentially devastating results.

It is also not the case that Heidegger is "against" technology, as some critics have maintained. Nor does he suggest any simple solution to the "danger" of pervasive enframing. He sees the current domination of technology and

---

[3] http://www.waterculture.org/uploads/Leopold_TheLandEthic.pdf

technological ways of thinking as part of the West's destiny. But at the same time the danger, that everything, including human beings, becomes a means to an end and thus infinitely manipulable, is real. At the end of his provocative essay, Heidegger reminds us that the Greek word "techne" from which technology is derived also means art and skill. The power which may save us from pervasive enframing is poesis or poetry. This may sound far-fetched, but Heidegger is reminding us that the Being of beings is revealed through poetry.

The Later 1970s and 1980s: The Concept of Sustainability; The Rise of Neo-liberalism

The later 1970s and 1980s saw contradictory phenomena: the emergence of large-scale transboundary environmental problems, requiring solutions across the borders and jurisdictions of nation-states, such as acid rain and ozone depletion, the establishment of a concept of sustainability, especially as enunciated in the 1987 UN Brundtland Commission Report, and the rise of neo-liberalism, pioneered in the USA and the UK, a political philosophy opposed to state intervention, relying on market mechanisms as the panacea to all problems. In his First Inaugural Ronald Reagan announced that "government is not the solution to our problems; government is the problem." At the same time it was concerted inter-governmental action which led to the Montreal Protocol which addressed and at least partially resolved the problem of the destruction of the ozone layer by CFC's used in aerosol sprays and refrigerators. In 1973 Frank Sherwood Rowland and Mario Molina of the University of California, Irvine, studied the impact of CFCs in the Earth's atmosphere and discovered that CFCs would break down in the middle of the stratosphere, releasing a chlorine atom which could interact with and break down large amounts of ozone (O_3), with potentially devastating results. After initial scientific doubts and strong denials on the part of industry, the problem was addressed and partially resolved with remarkable rapidity by the 1987 Montreal Protocol, banning the use of CFC's. One of the reasons this problem was resolved relatively quickly, academics have argued, was that less harmful and not especially costly substitutes for CFCs were available.

Another transboundary environmental issue of the time was acid rain, both in the USA and in Europe, caused mainly by power station emissions containing SO_2 and other corrosive agents which killed trees and acidified lakes. Once again initial denial, on the part of governments (notably the UK government) as well as industry, was followed by relatively swift action.

In 1979 the UN Economic Commission for Europe implemented the Convention on Long-Range Transboundary Pollution. Scrubbers were fitted to power station chimneys rendering their emissions less toxic.

The idea of sustainable development, or sustainability, especially as enunciated in the 1987 Brundtland Report, appeared for a while to offer the hope of reconciling the demands of development with the recognition of the severe and degrading pressures bearing down on the environment. In the well-known Brundtland formulation, sustainable development is "development which meets the needs of current generations without compromising the ability of future generations to meet their own needs." In practice, the concept has proved slippery, and, given the vagueness of the formulation especially as regards the specifics of the natural world, easy to interpret as little more than business as usual.

The 1980s in the West were dominated by the triumphant hegemony of free-market thinking, enthusiastically espoused by Ronald Reagan and Margaret Thatcher, among others. The decade ended with the fall of the Berlin Wall and the collapse of Communism in the Soviet Union and the former Soviet Bloc. Thatcher, however, surprised some by being the first prominent national leader to draw attention to the problem which would come to dominate thinking about the environment for the next decades, global warming, or as it came to be known, climate change.

Global Warming or Climate Change

The idea of the greenhouse effect, or the insulating effect of the earth's atmosphere associated with the heat-absorbing properties of certain atmospheric gases, especially carbon dioxide and water vapour, was not at all new. It had first been proposed by the French mathematician and physicist Joseph Fourier in articles written in 1824 and 1827. Fourier calculated that a body the size of the earth, at the distance the earth is from the sun, should be considerably cooler than it actually is, taking into account only the warming properties of solar radiation. Fourier suggested that the additional warming might be due to interstellar radiation, but also entertained the possibility of the atmosphere acting as an insulating blanket. He went even further, in a remarkably prescient paragraph: "The establishment and progress of human society, and the action of natural powers, may, in extensive regions, produce remarkable changes in the state of the surface, the distribution of the waters and the great movements of the air. Such effects, in the course of some centuries, must produce variations in the mean temperatures for such places."[4] This idea was confirmed by the work of the Irish-born physicist John Tyndall in the 1850s and 1860s, and then worked out in detail by the Swedish physicist Svante Arrhenius in the 1890s. Arrhenius was the first scientist to calculate how changes to the levels of CO_2 in the atmosphere could alter the surface temperature of the earth. Arrhenius formulated his greenhouse law as follows, in a paper published in April 1896 in the London, Edinburgh and Dublin Philosophical Magazine and Journal of Science: "if the quantity of carbonic acid (CO_2) increases in geometric progression, the augmentation of the temperature will increase nearly in arithmetic progression."[5] Interestingly, when Arrhenius came to write popular science books in later life, he suggested that human emissions of CO2 could have the positive effect of preventing another ice age.

Concern over possibly negative effects of global warming began to emerge in the 1950s, 1960s and 1970s. In 1979 the World Climate Conference of the World Meteorological Organisation (WMO) concluded that "it appears plausible that an increased amount of carbon dioxide in the atmosphere can contribute to a gradual warming of the lower atmosphere, especially at lower latitudes...It is possible that some effects on a regional and global scale may be detectable before the end of this century and become significant before the middle of the next century."

These concerns led the WMO, in partnership with UNEP, to set up in 1988 the Inter-governmental Panel on Climate Change (IPCC), the body which has become the main international focus and voice of concern on climate issues. At the Rio Earth Summit in 1992, the United Nations Framework Convention on Climate Change was signed by over 150 countries, with the objective of preventing "dangerous anthropogenic (i.e. human) interference of the climate system". Further discussions led to the Kyoto Protocol, signed in 1997, but only coming into force in 2005, setting emissions targets for developed countries which are binding under international law.

The hope was that the cardinally important issue and threat of climate change could be resolved in the same way that the problem of ozone depletion by CFCs had been resolved by the Montreal Protocol. The history has turned out very different.

The Kyoto Protocol has been weakened from the outset by the fact that the world's largest economy, the USA, has never signed up. One of the USA's reasons for not signing is that the Protocol does not include developing countries, including India and China, which in 2012 became the world's largest

---

[4] http://nsdl.library.cornell.edu/websites/wiki/index.php/PALE_ClassicArticles/archives/classic_articles/issue1_global_warming/n1-Fourier_1824corrected.pdf

[5] http://www.rsc.org/images/Arrhenius1896_tcm18-173546.pdf

emitter of CO_2. The Protocol has also been criticised by many other countries coming from different angles and positions. The setting of 1990 as the benchmark date from which increases and reductions in emissions were to be calculated has been criticised as being unduly favourable to European countries.

The most basic and telling criticism, perhaps, is that the Kyoto process has so far entirely failed to limit emissions of CO_2 and other greenhouse gases. These have continued to rise inexorably, as have levels of CO2 in the atmosphere as measured at Mauna Loa in Hawaii.[6] The rate of deforestation, another major cause of global warming, has also increased. Between 1990 and 2009, according to the Netherlands Environmental Assessment Agency, worldwide emissions of greenhouse gases rose by 40%.

The Kyoto Protocol also incorporated three market-based mechanisms, of which emissions trading is the most significant (the others are the Clean Development Mechanism and Joint Implementation). Under the emissions trading mechanism countries which have emissions units to spare are enabled to sell them to other countries which have exceeded their emissions quotas. The great hopes invested in these market mechanisms, so much in tune with the prevailing political Zeitgeist, have yet to be realised.

The Backlash: The Rise of Denialism

Possibly the most serious problem with the Kyoto process, or with this way of framing the question of mankind's relationship with the natural world, is that it has generated a backlash whose power threatens the whole environmental movement. You might argue that this was no different from other environmental or health issues, such as the over-use of pesticides, the depletion of the ozone layer, acid rain, or the establishing of a link between smoking and lung cancer. In all these cases, initial denial and counter-claims by industry, sometimes abetted or supported by public officials, were followed by acceptance of the science and remedial action.

Why has the story been so different in the case of the climate "debate"? Because cutting carbon emissions seems to imply not just a reconfiguration of one industry or industrial process but a sweeping change or suite of changes to a whole industrial economy reliant on fossil fuels, it is perhaps not surprising that the idea should meet with such concerted resistance, orchestrated by, among others, large oil companies deeply implicated in the political process. What is more surprising is that the credibility of climate science should have

---

[6] http://climate.nasa.gov/climate_resources/24/

been so comprehensively called into question. The reports issued successively by the IPCC in 1990, 1996, 2001, 2007 and 2013 have shown a steadily strengthening consensus on the part of climate scientists not only that climate change is occurring but it is very likely to have mainly anthropogenic causes: according to the 2013 IPCC report, "scientists are 95 % certain that humans are the 'dominant cause' of global warming since the 1950s.". But the strengthening of the scientific consensus has not led to a settling of the question, in the way that occurred with the ozone, acid rain and smoking and lung cancer questions. One of the world's largest and most respected media organisations, the BBC, has recently weakened its position on human-induced climate change.

The key lies in the degree of politicisation of the climate question, especially in the USA. An example of this is the change of position on climate science of the US Republican Party presidential candidate for the 2012 election, Mitt Romney. Romney's stated position in a response to the website ScienceDebate.org, sponsored by the National Academy of Sciences and other scientific organisations, according to the Washington Post, was "my best assessment is that the world is getting warmer, that human activity contributes to that warming, and that policymakers should therefore consider the risk of negative consequences." However, in a much-quoted speech given at the Republican National Convention in 2012, Romney mocked President Obama's promise to "slow the rise of the oceans and to heal the planet". According to Dana Nuccitelli in the Guardian newspaper "nearly 70 % of Republicans in Congress and 90 % of the party's congressional leadership deny the reality of human-caused global warming."[7]

Our Growing Estrangement from the Natural World

Global warming or climate change has come to dominate the environmental agenda to such an extent that other pressing environmental problems or questions are in danger of being neglected. Perhaps the most obvious is the question of the loss of biodiversity, or the depletion of the natural world. According to the biologist E.O. Wilson, "the sixth great extinction spasm of geological time is upon us, grace of mankind...In the world as a whole, extinction rates are already hundreds or thousands of times higher than before

[7] *The Guardian*, 3 July 2013.

the coming of man."[8] He estimates that "a fifth or more of the species of plants and animals could vanish or be doomed to early extinction by the year 2020, unless better efforts are made to save them."[9]

Concern and protection for biodiversity are enshrined in many international treaties and conventions (above all the UN Convention on Biological Diversity, but also, for instance, the Convention on International Trade in Endangered Species—CITES—and the Ramsar Convention on Wetlands of International Importance) but in practice the issue is not taken with the same seriousness as climate change. This could simply be because threats to other species are not considered as important as threats to humans, a legacy of the separation from nature which as we saw is particularly noticeable from the mid-seventeenth century onwards.

Another way of describing what is happening is that we are becoming progressively more estranged from nature or the natural world. In some contexts it even seems as if it is impossible to speak simply about nature: the word has now acquired a permanent halo, not in the least sacred, of inverted commas.

The American writer Richard Louv has written extensively about the estrangement of today's children from the natural world. He claims that "our society is teaching young people to avoid direct experience in nature."[10] He feels this is especially troubling given that "at the very moment that the bond is breaking between the young and the natural world, a growing body of research links our mental, physical and spiritual health directly to our association with nature."[11]

We live, it seems, in an increasingly artificial, electronic, digital and cybernetic world. In such a world, where the distinction between live and recorded birdsong, for instance, has been blurred, the question of the loss of biodiversity, even of the extinction of species, becomes somewhat remote from many people's concerns. As the naturalist Robert Michael Pyle has put it, what concern is the "extinction of a condor to a child who has never seen a wren?"[12]

The last decades have seen perhaps the most dramatic migration in human history, from the countryside into the city. In January 2012 it was reported that China's city-dwellers outnumbered its rural population for the first time in history. Even among city-dwellers, habits have changed, with children allowed far less time and space to roam freely than in previous ages.

---

[8] eowilsonfoundation.org.

[9] Ibid.

[10] Richard Louv, *Last Child in the Woods: Saving Our Children from Nature-Deficit Disorder* (Chapel Hill, North Carolina: Algonquin, 2005) pp. 2–3.

[11] Ibid.

[12] Ibid, p. 145.

In this context, it seems easier for people to imagine a largely artificial human existence. The natural world becomes merely a means of servicing human needs. Thus the still quite unfeasible idea of human colonisation of Mars is taken quite seriously in some quarters—for instance, by the physicist Stephen Hawking. For others, it has troubling connotations, by suggesting that having already depleted the resources of the mother planet, the Earth, the human species might move to ravage and despoil another planet.

Rewilding

Against this rather pessimistic vision of increasing artificialism and human disconnection from nature has arisen a counter-movement, with the aim of rewilding both nature and the human heart. This began in the US with the US Wildlands Project instigated by, among others, Dave Foreman, the founder of Earth First. Its manifesto aim is "to protect and restore the ecological richness and native biodiversity of North America through the establishment of a connected series of reserves". Foreman envisaged "gray wolf populations… continuous from New Mexico to Greenland" and "vast unbroken forests and flowing plains."

Rewilding has also come to Europe. There are calls for the reintroduction of top predators such as wolves and bear where these have become extinct. One of its earliest manifestations is the 5600 hectare Oostvaardersplassen reserve in the Netherlands, a wetland which has become a haven for wildlife including bird species such as bittern and spoonbill, wild Konik ponies, red deer and so on. There is a plan for Oostvaardersplassen to become part of a larger Oostvaardersland nature reserve, which in turn would be a part of Natura 2000, "the centrepiece of EU Nature and Conservation policy". Natura 2000 is "an EU-wide network of nature reserves" intended to ensure "the long-term survival of Europe's most valuable and threatened species and habitats."[13]

Natura 2000 is also a key element in the European Parliament Resolution on Wilderness in Europe passed by a large majority on 3 February 2009. This calls for the definition, development, protection and promotion of wilderness areas in Europe, in the context both of halting the decline of biodiversity and as part of a strategy to address climate change.[14] In 2013 the organisation

---

[13] http://ec.europa.eu/environment/nature/natura2000/index_en.htm
[14] http://www.europarl.europa.eu/sides/getDoc.do?pubRef=-//EP//TEXT+TA+P6-TA-2009-0034+0+DOC+XML+V0//EN

Wild Europe called for an assessment of progress on the European Parliament's Wilderness Resolution, calling it "a massive populist endorsement of the value of wilderness by Europe's directly elected representatives."[15]

Rewilding has also become the subject of a rapidly growing literary movement in the UK, led by writers such as Richard Mabey, George Monbiot (author of *Feral: Searching for Enchantment on the Frontiers of Rewilding*), Robert Macfarlane (whose third book, *The Wild Places*, recounts a journey around Britain's last remaining areas of wilderness) and Jay Griffiths (author of *Wild: An Elemental Journey*). Both Macfarlane and Griffiths link the loss of wilderness to the diminution of the human capacity for wonder.

---

[15] http://www.wildeurope.org/index.php/about-us/22-key-elements

6

Wider Attitudes to Environment

Beginning our story with a short history of Western cosmology, science and environmentalism is in no way intended to endorse a triumphalist view of the superiority and inevitable domination of Western ways of thinking and relating to nature. The geographer Ian Morris challenges the old idea that "two and a half thousand years ago…the Greeks created a unique culture of reason, inventiveness and freedom."[1] He explains the current dominance of Western thinking and technology more in terms of geography. All the same, that dominance is, for the time being, a fact of life.

But this does not mean that the prevailing Western way of relating to Nature is either desirable or sustainable. As Morris puts it, "at certain points the paradox of development creates tough ceilings that will yield only to truly transformative changes."[2] We seem to be at one of those points now, reluctant to admit that the enormous advances in wealth and comfort for the majority of human beings achieved by the scientific, industrial and informational revolutions may have potentially disastrous side-effects.

The western tradition certainly does not have a monopoly on attitudes to environment. In this section we examine some of the attitudes to the natural world found in the rich and ancient cultural traditions of Islam, Persia (Zoroastrianism), Asia (Buddhism, Hinduism, Jainism, Confucianism, Taoism), Africa (Yoruba, San), Pre-Columbian America and Australasia.

---

[1] Ian Morris, *Why the West Rules for Now* (London: Profile, 2010) p. 14.
[2] Ibid, p. 28.

© Springer International Publishing Switzerland 2017
H. Eyres, *Seeing Our Planet Whole: A Cultural and Ethical View of Earth Observation*, DOI 10.1007/978-3-319-40603-9_6

Islam, Persia

Islam is one of the three Abrahamic monotheistic faiths (with Judaism and Christianity) and shares many characteristics with them. The entire universe is the creation of God (Allah). It is Allah, like the Biblical Yahweh, who makes the waters flow, upholds the heavens, maintains the boundary between day and night and so forth.

A concept peculiar to Islam is khalifa. Mankind is God's khalifa, that is vice-regent or trustee. What is implied here is a stewardship; the Earth has been entrusted to mankind, and we are obligated to keep it safe and render account to God on the day of reckoning. The term for this accountability is akrah.

A central concept of Islam is tawheed or unity. It is incumbent on mankind as stewards of God to preserve the integrity of the earth, including non-human nature. In the Koran it is stated that "the world is green and beautiful and God has appointed you his stewards over it." Other Koranic verses speak of protection for trees and wildlife: "Whoever plants a tree and diligently looks after it until it matures and bears fruit is rewarded...If a Muslim plants a tree or sows a field and men and beasts and birds eat from it, all of it is charity on his part." Under shariah law animals are protected from cruelty, forests are conserved and the growth of cities is limited.

The period from the ninth to the twelfth century saw a magnificent flowering of Islamic mathematics and astronomy, both in the Middle East and in Spain, where Córdoba became the centre of Islamic culture and scholarship. The astronomer al-Battani calculated the length of the year to within two and a half minutes of the true figure. The Persian Muslim astronomer Abd al-Rahman al-Sufi catalogued 1022 stars, with far more precision than the Alexandrian Greek Ptolemy. This flowering withered after the reconquest of Córdoba in 1236. It is tantalizing to imagine what further discoveries the Islamic mathematicians and astronomers might have made in different circumstances.

Sufi Poetry

Sufism is a mystical branch of Islam, which spread widely through the Islamic world, from modern-day Morocco and Senegal in the west to Afghanistan and Pakistan in the east, starting around 850 AD. Under the influence of the greatest theorist of Sufism, Ibn al-'Arabi, "the earlier Sufi emphasis on public morals, supererogatory piety and the search for certain knowledge were subsumed into a grander vision of cosmic existence in which the infinite realizations of creative being now took center stage"[3]. Nature is often beautifully invoked in poetry by Sufi poets such as Rumi. A characteristic example is 'Birdsong':

> Birdsong brings relief
> To my longing
> I'm just as ecstatic as they are,
> But with nothing to say!
> Please universal soul, practice
> Some song or something through me!
> Tr. Coleman Barks

Zoroastrianism

Zoroastrianism has been called the first ecological religion. It stresses the preservation of the seven creations of God or Ahura Mazda, sky, water, earth, plants, animals, man and fire. In Also Sprach Zarathustra, Friedrich Nietzsche made Zoroaster/Zarathrustra, the main character of the book, the mouthpiece for his own philosophical ideas, some derived from Zoroastrianism. Zarathustra teaches the "Übersmensch", often but misleadingly translated "superman"; what he means by this is not a jack-booted fascist but a new kind of human, as far elevated above ordinary humanity as humanity is elevated above the apes. "The overman is the meaning of the Earth…I beseech you, my brothers, remain faithful to the Earth, and do not believe those who speak to you of otherworldly hopes!"[4] Nietzsche recognizes that following "the death of God", an enormous responsibility has been placed on man, which he may not be capable of discharging. We could interpret part of that responsibility as stewardship of the earth.

Taoism

Asian cultures have traditionally practised a more quietist and less interventionist approach in relation to the natural world than Western Christian and post-Christian ones.

It makes sense to start with the Chinese philosophy of Taoism, since this body of thought is concerned with mankind's relationship with nature.

---

[3] Nile Green, *Sufism: A Global History* (Chichester, West Sussex: Wiley-Blackwell, 2012) p.78.
[4] Friedrich Nietzsche, *Thus Spake Zarathustra*, Book 1, Zarathustra's Prologue, 3.

The Tao Te Ching stresses the preferability of Non-Doing (wu wei) to Doing. "The practice of Tao consists in daily diminishing," says Chapter 48. "Keep on diminishing and diminishing/Until you reach the state of Non-Ado."[5] But wu wei does not mean literally "do nothing" so much as "go with the flow" or "act spontaneously". By this kind of "doing nothing" as Lao Tzu says in a characteristic paradox, one can accomplish everything. Of course nothing could be less in the spirit of the Tao than such contemporary Chinese initiatives, guided by Western-influenced techno-rationality, as the Three Gorges Dam Project. This is one of the most extreme interventions in nature ever conceived; it has displaced millions, destroyed ancient cities and obliterated landscapes which inspired some of the greatest Chinese poetry.

Perhaps the most salient characteristics of the Tao are gentleness and subtlety. Its central message is that Nature has its own way (Tao means way); a good symbol for this is water, "which knows how to benefit all things without striving with them."[6] The implication here is that short-sighted human interference with this way is likely to lead to disaster. One of the most beautiful pairs of lines in the Tao goes as follows: "When the world is in possession of the Tao,/The galloping horses are led to fertilise the fields with their droppings."[7] This is an image of peace (the horses are not war horses), plenty (the fields are fertilised) and harmony. There is no conflict here between the "natural" and the human. The galloping horses, which have preserved their wild nature, are at the same time acting in harmony with human purposes.

Buddhism

Buddhism is generally more concerned with the cultivation of mindfulness and compassion than with relations between humans and the non-human world. However, according to the Dalai Lama, "the human attitude towards the environment should be gentle...if we exploit the natural environment in an extreme way, today we might get some benefit but in the long run we ourselves will suffer and other generations will suffer." Furthermore, the emphasis of Buddhism on the diminution of desire has implications in terms of a lighter

---

[5] Lao Tzu, *Tao Teh Ching* tr. John C.H. Wu (Boston, Massachusetts: Shambhala, 1961), p. 99.
[6] Ibid, p. 17.
[7] Ibid, p. 95.

footfall on the earth: "Buddhism calls for a modest concept of living: simplicity, frugality, and an emphasis on what is essential—in short a basic ethic of restraint."[8]

Wabi-sabi

Wabi-sabi is the quintessential, but almost undefinable Japanese aesthetic. It describes things of imperfect, evanescent beauty—of a certain roughness, and often on a small scale: things which might easily be overlooked, which do not shout from the rooftops but speak in a low, human tone.

"Wabi-sabi", according to Leonard Koren, "represents the exact opposite of the Western ideal of great beauty as something monumental, spectacular and enduring....Wabi-sabi is about the minor and the hidden, the tentative and the ephemeral: things so subtle...they are invisible to vulgar eyes."[9]

Wabi-sabi is not just an aesthetic, but also implies an ethics. The ethics of unpretentiousness leads to a lighter footprint on the earth.

Hinduism and Jainism

In Hindu teaching, all of creation is imbued with divinity; all living beings have a soul (atman). The greatest respect is therefore accorded to all living things, plants as well as animals. "If there is but one tree of flowers and fruit within a village," says the Mahabharata, "that place is worthy of your respect." The concept of ahimsa, shared by Hinduism, Jainism and Buddhism, means non-violence and enjoins Buddhists, Hindus and Jains not to cause harm to any creature. In practice this taken furthest by Jains, but animals (not merely cows) have also traditionally been treated with reverence by Hindus. The Yajurveda lays down that "no person should kill animals helpful to all. Rather, by serving them, one should attain happiness."[10]

The human race, although at the top of the evolutionary pyramid, is not seen as separate from the rest of nature. "What is needed today," according to His Excellency Dr Karan Singh, President of the Hindu Virat Saraj, in his 1986 Assisi Declaration, "is to remind ourselves that nature cannot be destroyed without mankind ultimately being destroyed itself."

---

[8] Padmasiri de Silva in Allan Hunt-Badiner (ed.) *Dharma Gaia: A Harvest of Essays in Buddhism and Ecology* (Berkeley, California: Parallax, 1990) p. 15.
[9] Leonard Koren, *Wabi-Sabi for Artists, Designers, Poets and Philosophers* (Point Reyes, California: Imperfect Publishing, 1994) p. 50.
[10] Yajurveda 13:47.

Egypt

Ancient Egyptian religion emerges with a multitude of gods, many taking the form of animals. Around 1400 BC a new sun religion briefly took hold under the Pharaoh who began his reign as Amenhotep IV before changing his name to Akhenaten, in honour of the sun-god Aten. Akhenaten dismantled the old temples and shrines of Thebes and constructed new temples dedicated to the sun-god. But this experiment ran up against resistance from the priesthood and after the deaths of Akhenaten and his son Tutenkhamun, Egypt returned to its former polytheistic ways.

Astronomy was highly regarded in ancient Egypt and star-watching priests had great prestige. The construction of the great pyramids of Gaza, as early as 3000 BC, bears witness to considerable mathematical knowledge.

Egyptian art as preserved in tomb paintings shows an exquisite sensitivity to the natural world, with birds, animals and plants depicted with extraordinary accuracy and beauty.

Yoruba

The Yoruba people are one of the most important ethnic groups in West Africa, especially south-west Nigeria and Benin. Yoruba culture and religion places stress on "God's divine lordship over the whole earth"[11]. The Yoruba idea of the environment is "all-embracing; the humans, animals, plants, and 'non-living beings' form the entire human society or community…[T]he tiniest of insects is regarded as having rights to life"[12]. Babalola also states that "sacred groves and wilderness are also seen as symbols of identity for all Yoruba people"[13]. She continues "sacred groves vary in size from a few hectares to a few kilometres. Protected by local residents as being the sacred residences of local deities and sites for religio-cultural rituals, they have served as valuable storehouse for biodiversity."[14]

---

[11] Fola D. Babalola, *Roles of and Threats to Yoruba Traditional Beliefs in Wilderness Conservation in Southwest Nigeria* (USDA Forest Service Proceedings RMRS-P-64.2011) http://www.fs.fed.us/rm/pubs/rmrs_p064/rmrs_p064_125_129.pdf
[12] Ibid.
[13] Ibid.
[14] Ibid.

San

The San (Bushmen) is the name given to the remnants of the hunter-gatherer people who once inhabited large parts of southern Africa. Now around 90,000 San remain, spread over South Africa, Namibia, Botswana, Angola, Zimbabwe and Zambia. Hunting is central to the San way of life, but, paradoxical as this may sound, great respect and even love is shown to animals in the act of killing them (often using poisoned arrows).

Pre-Columbian and Native American Cultures' Attitudes to Environment

A general idea has grown up of Native Americans as "spiritual ecologists"—peoples with a reverence for nature and dedicated to living in harmony with it. Some of this is derived from a speech supposedly delivered by Chief Seattle, a leader of the Duhamish tribe, in 1854. The speech as reported contains these words: "How can you buy or sell the sky—the warmth of the land? The idea is strange to us. Yet we do not own the freshness of the air or the sparkle of the water…Every part of this earth is sacred to my people…When the buffaloes are all slaughtered, the wild horses all tamed, the secret corners of the forest heavy with the scent of many men, and the views of the ripe hills blotted by talking wires, where is the thicket? Gone. Where is the eagle? Gone."[15] Unfortunately, much controversy surrounds the authenticity of the oration.

All the same, it seems clear from what we know of many Native American cultures that much less distinction was made between humans and non-human nature than in the Western tradition, that hunting carried with it certain mutual obligations, and that there was often a long-term concern for the sustainability of the way of life. Here for example is an extract from the Great Binding Law of the Iroquois people: "Look and listen for the welfare of the whole people and have always in view not only the present but also the coming generations, even those whose faces are yet beneath the surface of the ground—the unborn of the future Nation."[16]

In 1925, the Swiss psychoanalyst Carl Jung visited New Mexico and conversed with a Hopi elder named Mountain Lake. Mountain Lake told Jung that the Hopi believed that through their religious practices they helped the

---

[15] http://www.halcyon.com/arborhts/chiefsea.html
[16] http://www.constitution.org/cons/iroquois.htm

sun cross the sky every day. Here was a vision of human beings not just in harmony with the natural world but actively and intimately involved with its processes. Jung commented that "the idea, absurd to us, that a ritual act can magically affect the sun is no less irrational"[17] than the Christian religion.

Australasia

One of the distinctive beliefs of indigenous Australians is what Bruce Chatwin calls "…the labyrinth of invisible pathways which meander all over Australia and are known to Europeans as 'Dreaming-tracks' or 'Songlines'; to the Aboriginals as the 'Footprints of the Ancestors' or the 'Way of the Lore'."[18] Chatwin continues: "Aboriginal Creation myths tell of the legendary totemic being who wandered over the continent in the Dreamtime, singing out the name of everything that crossed their path—birds, animals, plants, rocks, waterholes—and so singing the world into existence."[19]

Conclusion

Western civilisations, especially since the Renaissance, the dawn of the scientific age and the Industrial Revolution, have proved immensely powerful and have spread their dominion (for considerable periods of time), their ideas and technologies over large tracts of the earth. Undoubtedly this has given rise to feelings of superiority and even arrogance. But recent trends in scholarship have done much to rehabilitate the ideas of other cultures. It is not only arrogant to assume that Western ideas, because of their power, are the only valid ones; it may also be foolish. The Hopi elder Mountain Lake confided to Carl Gustav Jung his misgivings about the white man and his attitude to nature. "We do not understand them…We think they are mad…They say they think with their heads. We think here"[20]; at this point Mountain Lake indicated his heart.

The ideas about nature and environment held by other cultures—for example, the Islamic concepts of stewardship and unity—are the result of centuries or millennia of sustainable living together with the natural world. It is less than two hundred and fifty years since the start of the Industrial civilisation and its sustainability for another quarter-millennium is far from assured.

---

[17] Carl Jung, *Memories, Dreams, Reflections* (London: Fontana, 1995) pp. 281–2.
[18] Bruce Chatwin, *The Songlines* (London: Jonathan Cape, 1987) p. 2.
[19] Ibid.
[20] Carl Jung, *Memories, Dreams, Reflections* (London: Fontana, 1995) p. 276.

7

The Slow Evolution of Environmental Ethics

Cultural attitudes to environment are always implicitly ethical: that is to say, they imply questions of how we should behave and act in relation to the natural world. As Carolyn Merchant has argued, envisaging the Earth as our mother is likely to imply a different way of interacting with the Earth than envisaging it as dead matter. It is possible to argue that the decisive break came in the sixteenth and seventeenth century, when Bacon, while still using the maternal metaphor, proposed that mankind should seek "in the womb of nature many secrets of excellent use", and Descartes made the distinction between res cogitans and res extensa.

Even so, we have seen from the preceding discussion how even in earlier times different ways of envisaging nature (for instance Democritus's atomism) might have different ethical implications. A specifically environmental ethics—that is to say an ethics which takes into consideration future generations and non-human species, or the whole of nature—has been slow to evolve. For example, the peer-reviewed journal Environmental Ethics studying the philosophical aspects of environmental problems was founded as recently as 1979.

In this section we will consider that slow evolution and argue, in agreement with Pope Francis in his recent encyclical, that a more profound and extensive recognition of environmental ethics is an urgent necessity in confronting our current crises, from global warming to the loss of biodiversity.

© Springer International Publishing Switzerland 2017
H. Eyres, *Seeing Our Planet Whole: A Cultural and Ethical View of Earth Observation*, DOI 10.1007/978-3-319-40603-9_7

The historical development of ethics can be seen in terms of widening circles of responsibility. In early human societies bonds of responsibility did not generally extend far beyond the group or tribe. Other groups or tribes would tend to be considered fair game or suitable for the pot. Even the ancient Athenians, often considered to have laid the foundations of western culture, considered non-Greek-speakers to be "barbarians". Of course, the extent to which humans have developed beyond this stage is highly questionable in the light of, just to take a few examples, the Crusades, the religious wars of the sixteenth and seventeenth centuries, the world wars and genocides of the twentieth century, and the ongoing conflicts of the twenty-first century.

Aristotelian Ethics

The first major treatise on ethics in the West is Aristotle's Nichomachean Ethics. Here Aristotle is primarily concerned with human well-being or flourishing (eudaimonia). Aristotle looks at what is required for "living well" (eu zen). Early on Aristotle makes the important point that flourishing needs to be considered not in the short term but over a whole lifetime. Aristotle also warns against the excessive pursuit of short-term pleasure. He looks at what he calls the dispositions of the soul which are conducive to flourishing and singles out the quality of "greatness of soul" or magnanimity. A magnanimous person will not harbor resentment and, in a crisis, will be prepared to risk his or her own life for a greater good. Another very strong emphasis in the Nichomachean Ethics is on friendship; friendship is discussed at greater length in the Nichomachean Ethics than any other topic (in Books VIII and IX) and it is clear that this quality reflects what is good and desirable not only on a personal level but also at the level of the community.

One might note first that the kind of subject Aristotle has in mind, implicitly rather than explicitly, is a free male citizen, not a woman or slave. Thus from the outset a large segment of society is excluded. Secondly flourishing or living well for Aristotle seems to be more about having a satisfying individual life than doing well by others. Somewhat set against this is the emphasis on friendship; Aristotle observes that friendship is one of the things humans need most. Here there is acknowledgement of the importance of others to an individual's flourishing. In another work, the Politics, Aristotle makes his famous remark that human beings are social or political animals (anthropos politikon zoon).

New Testament Environmental Ethics

New Testament ethics is revolutionary in its emphasis on altruism. In the parable of the Good Samaritan (Luke 10: 29–37) Jesus tells the story of a man travelling from Jerusalem to Jericho who is robbed, beaten, stripped of his clothes and left half-dead by the side the road. A priest and a Levite (member of an ancient Jewish priestly caste) approach but "pass by" on the other side of the road, doing nothing to help the man. Finally a Samaritan—an inhabitant of Samaria, a place despised by the Jews—comes along and offers succour to the injured man, tending to his wounds, and taking him to a nearby inn. Jesus' message in this parable is that ethical responsibility extends from the "in group" to the wider world.

In same gospel of Saint Luke, Jesus speaks of "the birds of the air" and says that "not one of them is forgotten before God" (Luke 12:6). Just after this Jesus draws attention to the "lilies of the field"; he remarks that "they toil not, they spin not; and yet I say unto you that Solomon in all his glory was not arrayed like one of these". Non-human nature is celebrated here and assigned value not for any usefulness but for its intrinsic essence and beauty and loveableness.

Saint Francis of Assisi

These ethical lines of thought, not dominant in the Pauline or Augustinian traditions, were especially revived by Saint Francis of Assisi. In the Canticle to Brother Sun and Sister Moon, composed around 1224, Francis sings the praises of the sun and moon, wind, air and water, and of "mother earth, who sustains and governs us, producing varied fruits with coloured flowers and herbs." Francis was famous for his love of birds and animals, and is depicted preaching to the birds in the cycle of frescoes attributed to Giotto in the Upper Church dedicated to him in Assisi.

Spinoza

Saint Francis's message of fraternity and solidarity with non-human nature stands as a lonely beacon in the panorama of succeeding centuries. The greatest treatise on ethics of the early modern era is Spinoza's radical Ethics, partly an attempt to free mankind from the dead hand of a personal deity and religious dogmatism. According to Spinoza, God is co-terminous with nature, envisaged as an active shaping force (natura naturans) and everything in the world has its own dignity and raison d'etre. But Spinoza did not go so far as St Francis in terms of fraternal feeling towards non-human nature.

Kant

Kant's ethical thinking and especially his formulations of the categorical imperative represent a major step forward in development of ethical thinking in the West. The three formulations, all from Grounding for the Metaphysics of Morals, run as follows: (1) Act only according to that maxim whereby you can at the same time will that it should become a universal law without contradiction. (2) Act in such a way that you treat humanity, whether in your own person or in the person of any other, never merely as a means to an end, but always at the same time as an end. (3) Therefore, every rational being must so act as if he were through his own maxim always a legislating member in the universal kingdom of ends.[1]

Kant argued that the grounding of morals lies nowhere else but in human reason; to be more specific, in what he called "pure practical reason". Human beings enjoy free will and at the same time are called upon to act in accordance with universal moral laws, that is to say under the "categorical imperative". For Kant, moral laws make no logical sense unless they can be universalized. And these moral laws lead inexorably to the conclusion that no human being should ever under any circumstances treat any other human being merely as a means to an end. In the Metaphysics of Morals he extends his ethical consideration to non-human nature, by prohibiting cruelty to animals. However, he derived this prohibition from the argument that such cruelty would tend to deaden compassion in the human subject, rather than by granting the status of ends in themselves to animals.

Self and Other: Hegel, Merleau-Ponty, Levinas

Noble as it is, Kant's categorical imperative does not address the sheer thorniness of the engagement of the self with the other. This is one of the most important themes in the further development of psychological and ethical

---

[1] See http://www.earlymoderntexts.com/assets/pdfs/kant1785.pdf

thinking in the nineteenth, twentieth and twenty-first centuries. In what sense can we know or do we need the Other? Is our relationship with the Other doomed to be confrontational or oppositional? Are we out merely to appropriate what we can of the Other and the Other's thinking, or can we engage in productive and mutually enhancing dialogue?

In an obvious sense we are all self-centred. As babies we are unavoidably taken up with our own needs. The arrival of younger siblings is not always greeted with undiluted pleasure. At the same time, as both philosophers and psychologists have argued from Hegel onwards, we are always from the beginning in communication with the Other.

Here we can make a parallel with cosmological thinking; geocentric theories can be seen as an obvious extension of "self-centredness". Of course it makes sense to human beings, proud of their position at the apex of the chain of being (below God and the angels!), to assume that they are at the centre of the universe. In some ways Copernicus' great discovery, confirmed by Galileo, has yet to be fully assimilated on a psychological level. Earth Observation can also be seen as a development away from self-centredness; instead of looking out from our personal point of view we are looking back at ourselves, our planet, and the impact we are having on it from a distance.

Hegel

At the beginning of his great treatise The Phenomenology of Spirit Hegel writes that "The significance of that 'absolute' commandment, 'know thyself'…is not to promote mere self-knowledge in respect of the particular capacities, character, propensities, and foibles of the single self."[2] According to Frances Berenson, "Hegel writes that the Other Self is the only adequate mirror of my own self-conscious self; the subject can only see itself when what it sees is another self-consciousness."[3] But if Hegel acknowledges that the self can only be known in relation to the other, the example he gives of this process, the Master-Slave dialectic, does not seem particularly encouraging. In the first stage of this, according to Berenson, "each person only fully recognizes his own conscious being and belittles any claim to other self-consciousness. Here each self seeks to assert his own self-consciousness even at

---

[2] G.W.F. Hegel, *The Phenomenology of Spirit*, Introduction, 377 https://www.marxists.org/reference/archive/hegel/works/sp/suintrod.htm
[3] Frances Berenson, 'Hegel on Others and the Self', *Philosophy* vol. 57 no. 219.

the price of destroying the life and self-consciousness of others, even though this risks his own existence."[4]

In Hegel's defence, we should acknowledge that this is only an early stage of the dialectic. Hegel does not propose the Master-Slave relationship as a satisfactory one. In a passage which has given rise to multiple interpretations, Hegel suggests that through the crucible of the unsatisfactory Master-Slave relationship a more rewarding one of mutual recognition and self-knowledge may be achieved.

Merleau-Ponty

In the twentieth century phenomenologists and psychoanalysts have greatly developed and refined Hegel's suggestive "myth" of Self and Other. The phenomenologist and child psychologist Maurice Merleau-Ponty "affirms an interdependence of self and other that involves these categories overlapping and intertwining with one another, but without ever being reduced to each other."[5] The child only comes to recognize itself through the mirroring which occurs in its closest and earliest relationships, especially with the mother. Merleau-Ponty differs from Hegel in stressing the importance of embodiment rather than abstract intelligence. Merleau-Ponty with his concern for the bodily nature of things (not only human beings) can also be seen as a precursor of ecophenomenology: his last working note reads as follows: "Nature as the other side of humanity (as flesh, nowise as "matter")."[6]

Levinas: Responsibility for the Other

Merleau-Ponty's main concern is with perception rather than ethics. However, his criticism of the objectifying gaze of science, which turns everything into objects, has ethical implications. If a person looks at someone else with an objectifying gaze, even with the best scientific intentions, he or she turns that other into a thing, a mere object of study, not a person. The objectifying gaze is as far as can imagined from the face-to-face encounter with the Other which is central to ethics as championed by the greatest twentieth century

---

[4] Ibid.

[5] http://www.iep.utm.edu/merleau/

[6] Maurice Merleau-Ponty, *The Visible and the Invisible*, p. 274 https://monoskop.org/images/8/80/Merleau_Ponty_Maurice_The_Visible_and_the_Invisible_1968.pdf

philosopher of ethics, Emmanuel Levinas. For Levinas ethics is fundamentally concerned with responsibility, and particularly with responsibility for and towards the Other.

There is nothing abstract about Levinas's radical vision of ethics. Of Lithuanian-Jewish ancestry, he was born in Kaunas, moving to France to study at Strasbourg University in 1924. Having enlisted in the French Army at the start of World War Two, he spent much of the war in a German prisoner of war camp for Jewish prisoners. His father and brothers were murdered in Lithuania by the SS while his mother-in-law also disappeared during the war.

It was out of this context of the mass-murder of millions of human beings, treated purely as objects by their killers and by the bureaucracy which planned their deaths, that Levinas developed his notions of ethics as first philosophy and of the primacy of the Other.

The ethical black hole of the Holocaust brought forth a radical new vision which challenged the whole of Western philosophy's concern with being (ontology) and knowledge (epistemology). For Levinas ethics comes before philosophy. In fact ethics, as responsibility, even comes before being. Subjectivity is constituted out of responsibility: I only "am", as a full human being, insofar as I am responsible to and for the other. "The tie with the Other is knotted only as responsibility, this moreover, whether accepted or refused, whether knowing or not knowing how to assume it, whether able or unable to something concrete for the Other. To say: here I am [me voici]. To do something for the Other. To give. To be human spirit, that's it."[7]

Levinas's apparently extreme emphasis on responsibility for the Other contains no explicit reference to the environment. However, if the notion of the Other is extended to include future generations and non-human nature, Levinasian ethics points the way towards a very strong version of environmental ethics. To be fully human, for Levinas, is to be responsible for and on the side of the Other. This challenges our customary self-centredness, encouraged by the kind of economics and politics which promotes individual satisfaction, choice and consumption, often assumed to exist in a kind of bubble, as if individuals and their choices did not impinge on others – or, to use the language of environmental economics, as if there were no externalities. Extending Levinasian ethics in an environmental direction would imply a responsibility to and for future generations and non-human nature. But this responsibility should not be imagined as a sort of extra duty, but seen as a constitutive part of our humanity and as giving meaning to our lives.

[7] Emmanuel Levinas, *Ethics and Infinity* (Pittsburgh, Pennsylvania: Duquesne University Press, 1985) p. 97.

Pope Francis: Laudato Si

This is very close to the vision of Pope Francis in his encyclical Laudato Si (24 May, 2015). Towards the end of the letter Pope Francis speaks both of the spiritual shortcomings of the current consumer society, based on "the mere amassing of things and pleasures", which "are not enough to give meaning and joy to the human heart", and the importance of environmental education, not merely to provide information but to reorient people towards an ecologically responsible way of living. Environmental education "seeks... to restore the various levels of ecological equilibrium, establishing harmony within ourselves, with others, with nature and with God. Environmental education should facilitate making the leap towards the transcendent which gives ecological ethics its deepest meaning."

The Globalization of Ethics

For centuries geo-politics in the West was governed by the principles laid down in the Treaty of Westphalia which put an end to the Thirty Years' War in the Holy Roman Empire and the Eighty Years' War between Spain and the Dutch Republic. The Westphalian "order" was based on the sovereignty of states, whose aggressive tendencies might be held in check by a "balance of power". What went on within each state was not the business of other states.

The Westphalian Order was challenged by the events of World War Two and especially the genocide of the Jewish people (together with Romany gypsies and other groups) under the Nazi regime – though there had already been an attempt to set up some kind of international tribunal under the League of Nations following World War One. In the aftermath of the Second World War and the revelations of the horrors of the holocaust and other war crimes, International Military Tribunals for war crimes were set up in Nuremberg and in Tokyo.

The development of an International Criminal Court was slow and tortuous but was eventually achieved in 2002 following further International Military Tribunals to prosecute war crimes in the former Yugoslavia and in Rwanda.

The sheer lengthiness and difficulty of this process, together with the ongoing horrific war crimes it is designed to address, may demonstrate the difficulties surrounding the attempt to extend ethical responsibility beyond the boundaries of nation states to the whole human "family". The proposal that one might extend ethical responsibility still further, beyond the human family to other species, and into the future, might appear fanciful in the extreme. But this is the essential premise of environmental ethics.

Intra- and Inter-Generational Justice

One of the issues which has exercised and bedevilled international negotiations around climate change and the limiting of greenhouse gas emissions concerns the unequal responsibility for climate change and the unequal impacts it has and is likely to have. Certain countries—those of the rich north—are far more responsible for greenhouse gas emissions and the rise of CO2 concentrations than others. This is why in the Kyoto process emissions targets and limits were initially imposed only on developed countries. At the same time, the effects of climate change are already being felt, and will in the future be felt, disproportionately by certain developing countries which suffer from droughts, severe floods, vulnerability to sea level rise and so on. In Bangladesh, for instance, it is reckoned that "millions of people are already being displaced by natural hazards and that many millions more will be displaced in the future as a result of the increased frequency and severity of natural hazards due to climate change."[8]

These complicated ethical issues are also hard for democracies to address, since democratic polities are constituted so as to be responsive and responsible to their electorates, not to all people and entities which may be affected by a policy or set of policies.

Intergenerational justice issues—concerning the effect of climate change on generations still to be born—are even more complicated than intragenerational ones. In standard economics the solution would be via discounting. But here we need to discuss the ethics of discounting.

The Ethics of Discounting

Discounting is based on the assumption that capital and resources invested now, for instance in the fight against climate change, will become vastly more valuable in the future, because of predicted steady per capita consumption growth at compound interest. Even a relatively low estimated growth rate such as the 1.3% used by Nicholas Stern in his calculation of the discount rate used in the Stern Report carries the assumption that consumption would triple by 2100.[9] This justifies the idea that the interests of future generations can be "discounted" against our own, as they will be vastly richer than us (partly as a result of investments we make).

---

[8] http://displacementsolutions.org/wp-content/uploads/DS-Climate-Dsplacement-in-Bangladesh-Report-LOW-RES-FOR-WEB.pdf
[9] Dale Jamieson, *Reason in a Dark Time* (Oxford: Oxford University Press, 2014) p. 118.

Famously, Stern's "low" discount rate gave rise to a furious disagreement between the British economist and the American William Nordhaus, an advocate of a much higher rate of 5%, which included the accusation by Nordhaus that the Stern Report was "stoking the dying embers of the British Empire".[10] In fact Stern and Nordhaus disagree over much more than the discount rate. Stern takes what has been called a "prescriptive" approach, explicitly bringing in ethical considerations, while Nordhaus operates under the "descriptive" approach, which asks "What choices involving trade-offs across time do people actually make?"[11]

There are in fact much more profound and wide-ranging reasons to doubt that an economic mechanism such as discounting can adequately capture the future costs and benefits of climate change. One argument is that the effects of climate change are highly unpredictable and non-linear and contain elements such as tipping-points and feedback mechanisms which are impossible to calculate with the exactitude needed for neat economic models such as the Integrated Assessment Models (IAMs) used by both Stern and Nordhaus. Another argument is that such economistic valuing, without finer distinction, could equalize say the costs of deaths from an epidemic with the resulting increased profits of a pharmaceutical company.[12] A further argument is that there are certain values which are hard if not impossible to put a figure on (the cost of a species becoming extinct, or the loss of a particular ancient forest, for instance).

Climate change carries the very real risk and possibility of catastrophic impacts. For this reason, as Jamieson argues, there has been a move away from the discounting methods used by both Stern and Nordhaus and towards the idea of investing in climate change mitigation as a kind of insurance policy, taken out on behalf of the future. The pioneer here was the climate scientist Stephen Schneider, who once wrote that "we buy fire insurance for our house and health insurance for our bodies. We need planetary sustainability insurance."[13]

Beyond Legality and Instrumentality

To understand the force of environmental ethics, one must go beyond notions of mere legality or instrumentality. Legal norms, though vital for the functioning of a good society, must clearly be distinguished from ethical norms.

---

[10] Ibid, p. 111.
[11] Ibid, p. 122.
[12] Ibid, p. 142.
[13] Ibid, p. 140.

Legal norms tend to concentrate on what is permitted, not what is good. They tend to be relatively weak. Ethical norms are stronger. In their strongest form, as evinced in Christian ethics and in Levinasian ethics, they demand the placing of the interests of the Other either on an equal footing or above one's own interest. When asked which is the great commandment, in Matthew 22:35–40, Jesus replies "Thou shalt love the Lord thy God with all thy heart, and with all thy soul, and with all thy mind. This is the first and great commandment. And the second is like unto it, Thou shalt love thy neighbour as thyself. On these two commandments hang all the Law and the Prophets." Levinas seems to go even further and to define ethical relation, according to Philippe Nemo, as "the infinite responsibility of being-for-the-other *before* oneself."[14]

A currently fashionable version of environmentalism, termed by the environmental writer Paul Kingsnorth "neo-environmentalism", as a parallel to neo-conservatism, holds that conservation should not be for the sake of nature per se, but only for those aspects of nature which are benefit to humans. This clearly falls far short of a Levinasian conception of ethics, as being essentially concerned with responsibility for the Other. In neo-environmentalism there is no responsibility for nature as Other, including other species.

Speciesism: Peter Singer

A forceful articulation of the interests of other species has been made by the Australian utilitarian moral philosopher Peter Singer. In his 1975 book *Animal Liberation*, Singer argues that the distinction between animals and humans is purely arbitrary, given that the difference between a human and a gorilla, say, is vastly smaller than that between a gorilla and an oyster. In terms of intelligence, some animals can display intelligence comparable to or superior to that of a human child, or a mentally handicapped person. Thus using intelligence as the distinction between humans and animals does not work. Singer calls the prejudice in favour of humans over animals "speciesism" (a term originally coined by the psychologist Richard D. Ryder), an obvious reference to other forms of prejudice and discrimination such as racism and sexism.

---

[14] Emmanuel Levinas, *Ethics and Infinity* (Pittsburgh, Pennsylvania: Duquesne Univerity Press, 1985) p. 12.

Animal Rights

Singer does not speak about animal rights but rather about animal interests. He argues for a protection of animal interests. Other philosophers go considerably further. Tom Regan argues that (some) animals are "the experiencing subjects of a life", with beliefs, memories, and a sense of their own future, who must be treated as ends in themselves, not means to an end.[15] Obviously this raises as many questions as it answers. Where are we to draw the line, between, say orang-utans and mosquitoes? At the same time, we find ourselves back in the territory of some ancient religious beliefs, such as that of the Jains in India, who hold all life to be sacred.

Conclusion

Ethical implications are inescapable in all of this. There is a temptation to think of climate change and other environmental problems as purely technical or at the most economic problems. However, it is in fact partly because climate change will impact on future generations, on other species and on the some of the basic ecosystem capacities of the planet that we are confronted with inescapably ethical choices. If ethics is concerned with our responsibility to the Other, and the Other includes other species and future generations, then we cannot simply duck these choices, however awkward or costly. Carrying on with "business as usual" is not the neutral, default position many imagine it to be. Satellite Earth Observation, by providing reliable, comprehensive and continuous monitoring of the changes actually occurring, for instance in sea ice extent, sea level rise and sea temperature, forestation and other aspects of the physical environment, provides us with essential information on which to base these moral choices.

---

[15] http://www.animal-rights-library.com/texts-m/regan03.htm

8

A Short History of Earth Observation

The Earliest Times; Balloons; Aerial Photography

The history of Earth Observation no doubt begins with the views over the earth revealed to intrepid early mountain-climbers and fantasies of the still greater vistas available to soaring hawks, eagles and vultures.

It was only in the eighteenth century, with the development of hot-air balloons, that men and women gained the ability to ascend significant distances into the aether, and at the same time to begin to make scientific measurements of and from the upper air. Kites had already been used, by Benjamin Franklin among others, to carry scientific instruments into the atmosphere; one of Franklin's experiments proved that lightning was electricity.

The Boston physician John Jeffries, who made two balloon flights with the French balloonist Jean Pierre Blanchard, including the very hazardous first Channel crossing by balloon in January 1785, is credited with being the first person to make meteorological measurements from a balloon.

The invention of photography combined in the nineteenth century with the technology of ballooning to make possible the first aerial photographs, including those of Nadar (Gaspar Felix Tournachon), taken from his Le Géant balloon in 1858. Nadar inspired the character of Michel Ardan in Jules Verne's 1865 novel From the Earth to the Moon about three adventurers being launched into space by a cannon.

© Springer International Publishing Switzerland 2017
H. Eyres, *Seeing Our Planet Whole: A Cultural and Ethical View of Earth Observation*, DOI 10.1007/978-3-319-40603-9_8

The next major development was the invention of aircraft in the early years of the twentieth century. When the First World War broke out in 1914, the use of aerial photography gained a new relevance in facilitating much more accurate intelligence about enemy forces and positions. The development of rockets and jet aircraft allowed aerial photographs to be taken from greater altitude. In 1929 the great American rocket pioneer Robert H.Goddard launched a rocket with a camera, able to take photographs of clouds.[1] Then in the late 1940's, the first photographs of earth and its cloud patterns from the edge of space were taken by research rockets.

The First Satellites

The great breakthrough, however, was the successful launch of the first artificial earth satellite, Sputnik 1 by the USSR in 1957. The Russians had stolen a march on the Americans, despite the announcement by the US in 1955 that it would launch a scientific Earth satellite during the International Geophysical Year of 1957–1958. Very quickly the potential for satellite Earth Observation was spotted and used. Much of the impetus came from the "space race" between the USSR and the US in the context of the Cold War.

The first US satellite Explorer 1 was launched a few months later on January 31, 1958. That year also saw the creation of the National Aeronautics and Space Administration—NASA—and the momentous statement by the President's Science Advisory Committee outlining the uses of an orbiting satellite: "(1) It can sample the strange new environment through which it moves; (2) it can look down and see the earth as it has never been seen before; and (3) it can look out into the universe and record information that can never reach the earth's surface because of the intervening atmosphere."

The US's first satellite with meteorological instruments, Vanguard 2, was launched on 17 February 1959 and was followed by Explorer 6, launched on 7 August of the same year, which returned the first satellite Earth photo. Explorer 7, launched in October 1959, carried an impressive array of instruments including the improved radiometer devised by University of Wisconsin scientist Verner Suomi. This took the measurements of Earth's radiation from space which mark the beginning of satellite climate studies. Suomi also devised a spin-stabilised camera which was attached to NASA's ATS 1 communications satellite and which was able to produce full-disk images allowing the monitoring of complete weather patterns.

---

[1] http://www.nasa.gov/centers/goddard/about/history/dr_goddard.html

On 1 April 1960, the first satellite entirely dedicated to satellite meteorology, the Television and Infrared Satellite (TIROS) 1 was launched. The far-reaching result was that for the first time humankind could view the Earth and its weather systems as a whole. Satellite observation has transformed "our perception of the Earth from a set of distant, isolated continents to an integrated system of land, ocean, atmosphere and living things."[2]

The development of satellite meteorology is one of the great and least trumpeted achievements of modern science. The scope and sophistication of instruments has steadily increased, with developments such as Automatic Picture Transmission (introduced on TIROS 8 launched in 1962), the infrared spectrometer and infrared interferometer designed to provide atmosphere soundings and carried on Nimbus 3 launched in 1969, and even more important the development and launch of geostationary metsats starting with Synchronous Meteorological Satellite (SMS) 1 launched in 1974, followed by GOES 1 launched in 1975 and ESA's Meteosat launched in 1978.

NASA's seven Nimbus missions, launched into sun-synchronous polar orbit between 1964 and 1978, went beyond meteorology to the mapping of gases in the atmosphere, the temperature of the ocean, the tracking of ocean currents, observation of the ozone layer, the mapping of sea ice and measurement of atmospheric temperature.

The Global Weather Experiment

At the same time as Nimbus 7, the Global Atmospheric Research Programme under the auspices of the WMO and the ICSU launched the Global Weather Experiment, the largest international scientific experiment ever undertaken. The objectives included "(1) Understanding atmospheric motion for the development of more realistic models for weather prediction. (2) Assessing the limit of predictability of weather systems. (3) Designing an optimum composite meteorological observing system for routine weather prediction of the larger-scale features of the general circulation. (4) Investigating the physical mechanisms underlying climate fluctuations and to develop and test appropriate climate models." This was done using "the World Weather Watch (WWW) surface/upper-air network and voluntary observing ships, commercial aircraft, polar orbiting and geostationary satellites, drifting meteorological buoys mainly in the southern hemispheric ocean".[3]

---

[2] http://climate.geog.udel.edu/~tracyd/geog674/geog674_history.html
[3] http://gcmd.nasa.gov

Satellite metsats have been operational since the 1960s, with spectacular results. Since that time, there have been no undetected tropical cyclones anywhere on Earth, though even with today's extremely sophisticated instruments the precise path of cyclones remains difficult to predict, as was the case with cyclone Pam which devastated Vanuatu in March 2015.

From Metsats to Earth Observation

Satellites, combined with ever more sophisticated and powerful computers and information technology, turned out to be able to offer continuous and global monitoring of weather systems; to measure atmospheric pressure, temperature and moisture changes and other variables. The services they provided, including the steadily improving ability to forecast weather, also proved attractive, especially to national meteorological services, as well as to some extent the private sector.

EUMETSAT

The USA had led the way with meteorological and Earth Observation Satellites, especially the missions launched by NASA (Explorer, Nimbus) and NOAA, and the successful ERTS (later Landsat) programme for land remote sensing with extremely high resolution launched in 1972. Europe stepped up to the plate in the 1980s. The European Organisation for the Exploitation of Meteorological Satellites—EUMETSAT—based in Darmstadt, Germany, was established through an international convention which opened for signature in 1983 and came into force in 1986. It now has 30 members, including all the member states of the EU bar Malta plus Norway, Switzerland and Turkey; it is funded by national contributions of member states which are proportional to gross national income. EUMETSAT runs a comprehensive fleet of weather satellites flying on two different orbits, geostationary and polar sun-synchronous. The geo-stationary satellites flying at 36000 km above the equator provide continuous monitoring; polar-orbiting satellites such as Metop and Jason flying at much lower altitude provide more granular data although with less frequent global coverage.

The economic and social benefits of accurate weather forecasting, in such areas as agriculture, transport, tourism and energy supply as well as in the avoidance of disasters such as floods, are extremely significant: EUMETSAT estimates the value of its services in relation to protection from floods, storms and other weather events to the European economy at up to £5.4 billion per year.

EUMETSAT itself has provided a model for the expansion from mete-orological satellite observation to broader systems of Earth Observation. EUMETSAT's remit now includes climate and environment as well as weather monitoring. According to the director Alain Ratier, "satellite observa-tions are relevant to climate monitoring. Over the last 30 years meteorological satellites like Meteosat have indeed accumulated unique records of our chang-ing climate."[4] Satellites clearly have advantages over in situ measuring devices when it comes to monitoring changes occurring over vast and inaccessible regions such as the oceans, the polar icecaps, remote glaciers and areas of roadless rain forest. All of these are of especial significance when it comes to climate change. In terms of international co-operation, a notable achievement has been the successful collaboration between EUMETSAT and NOAA: the two agencies agreed to provide mutual back-up support in 1993 and in 1995 "signed an agreement that acknowledged and regulated the reciprocity of their exchange of data."[5]

Enter the EU: Towards GMES/Copernicus

A significant development occurred in May 1998 during the annual User Seminar of the Space Applications Institute of the Joint Research Centre of the EU in Baveno, Italy. Under the instigation of the then Director-General of the JRC, Herbert Allgeier, the so-called Baveno Manifesto was launched, reaffirming the will of a number of agencies and organisations involved in space activities to "contribute to a common European vision and strategy towards global environmental monitoring." Furthermore, the Baveno mani-festo underscored "the need for Europe to act if it wants to maintain a leading position in the monitoring techniques for a better understanding of environ-mental changes, resource depletion and environmental risks at a global scale."

Baveno and Kyoto

An important part of the context for the Baveno initiative was the UN Convention on Climate Change and the Kyoto Protocol. "Supporting the commitments made by the [European] Community in the context of the 1997 Kyoto Protocol on the reduction of greenhouses gases," states the Manifesto, "represents a priority of today…The Kyoto Protocol calls…for monitoring

---

[4] Conversation with the author, 7 March 2013.
[5] Arne Lahcen, ESPI Report 46, p. 5.

global progress towards reducing greenhouse gas emissions and enhancing carbon sinks." The establishment of a European "global information service", it continues, "would represent a significant European contribution to the implementation of the Kyoto Protocol."

The Manifesto also underlines the vital contribution made by Earth Observation satellites to this global monitoring process, as part of the "stewardship of our global environment", and cites the success of satellite weather monitoring systems. From the beginning, the Baveno initiative linked environmental monitoring and security, seeing presciently that "environmental problems…may, firstly, endanger the security of both individuals and nations and, secondly, lead to international conflict."

A year after the May 19, 1998 Baveno meeting which led to the issuing of the Manifesto, a second meeting was held in the same Italian town which resulted in a paper setting out the vision for a new European initiative more concretely. Once again this paper stressed the dangers of "environmental degradation and natural resource depletion" and the importance of "safeguard[ing] the interests of future generations through international environmental treaties, protocols and conventions."

European Responsibility

The 1998 Baveno Manifesto and the 1999 paper are calls for action based on values, principles and responsibilities seen as being at the heart of the whole European project. This is made explicit in the 1999 paper, where it is stated that "understanding the climate and environment is a task for all nations but one in which the more advanced and wealthy are expected to lead. If Europe aspires to such a mantle then it must play a leading role in this field." According to Herbert Allgeier the creation of what would become GMES was part of "the constitution of Europe".[6] Mr Allgeier also made it clear that "European involvement in space needs the political dimension of the EU otherwise it has no future. This political EU has to spell out ambitions and limits of ambitions and define programmes which deliver those ambitions".[7]

Herbert Allgeier's vision came to fruition with the creation of GMES, later renamed Copernicus, as a world-leading system of global monitoring for environment and security, co-ordinated and managed by the EC with the

---

[6] Conversation with the author, 21 January 2013.
[7] Ibid.

space observation infrastructure performed under the aegis of ESA, using data from a wide variety of sources including the new generation of Sentinel satellites, of which the first was launched in 2014.

Via its C-band Synthetic Aperture Radar, Sentinel 1 "provide[s] an all-weather, day-and-night supply of imagery of Earth's surface."[8] The mission will also "provide services relating to monitoring of Arctic sea-ice extent, routine sea-ice mapping, monitoring land surface for motion risks, mapping for forest, water and soil management and mapping to support humanitarian aid and crisis situations."[9]

Between 14 and 16 August 2015 radar images from Sentinel 1-A captured the Jakobshavn glacier in Western Greenland—the world's fastest-moving glacier—before and after the calving event in which it "shed a chunk of ice measuring around 12.5 cubic kilometres."[10] The implications of increased ice melting in Greenland are of course extremely significant: it is estimated that the melting of the Greenland icecap would raise sea levels worldwide by approximately 6 metres. Coastal and estuarial cities such as New York, Shanghai, Hong Kong, Rio de Janeiro, Mumbai, Dhaka and London would be seriously threatened and might become uninhabitable.[11]

Sentinel 2-A was launched in June 2015. The Sentinel 2 mission "will mainly provide information for agricultural and forestry practices and for helping manage food security." It will also "monitor the world's forests and provide information on pollution in lakes and coastal waters."

But the scope is actually wider, and goes beyond management to environmental stewardship, and more specifically to include the challenge of monitoring biodiversity from space. This is of crucial importance given the fact that "global biodiversity loss is intensifying",[12] or to put it in less specialised language, that we are steadily and progressively destroying and degrading the natural world. According to Professor Andrew Skidmore, "as conservation and remote-sensing communities join forces, biodiversity can be monitored on a global scale with the assistance of satellites."[13]

---

[8] http://www.esa.int/Our_Activities/Observing_the_Earth/Copernicus/Sentinel-1/Introducing_Sentinel-1
[9] Ibid.
[10] http://www.esa.int/Our_Activities/Observing_the_Earth/Copernicus/Sentinel-1/Chasing_ice
[11] https://nsidc.org/cryosphere/quickfacts/icesheets.html
[12] http://www.nature.com/news/environmental-science-agree-on-biodiversity-metrics-to-track-from-space-1.18009)
[13] http://www.esa.int/Our_Activities/Observing_the_Earth/Help_wanted_on_tracking_biodiversity_from_space/(print)

Copernicus is a data resource covering six thematic areas, land, marine, atmosphere, climate change, emergency management and security. After difficult negotiations, the funding arrangements keeping Copernicus within the Multi-Annual Financial Framework (MFF) of the EU were secured in 2013. This was extremely important, as removing GMES from the MFF, with financial negotiations having to be made on a multi-lateral basis, would undoubtedly have led to delays and uncertainties which might have threatened the project's continuity. Copernicus moved into the operational phase in 2014.

By using the language of "stewardship" and speaking of "future generations" the Baveno manifesto and papers which formed the basis of GMES/Copernicus set the issue of environmental monitoring within a framework which is not merely economic and political but also ethical. The highlighting of the context of Kyoto shows that concerns about human-induced climate change, caused by the build-up of greenhouse gases, are at the heart of this initiative.

Values and Principles

This in turn is connected with the foundational values and principles of the EU. Here a short digression is in order. The EU is often seen as a primarily economic entity. Indeed, its origins as a coal and steel union give every appearance of being economic. But this pragmatic appearance was always something of a facade masking broader and deeper aims and principles.

As the former EU Commissioner and Director-General of the WTO Peter Sutherland has argued, the founding fathers of the EU were always motivated by philosophical and ethical considerations as well as economic ones. "The founding fathers of the EU were explicitly philosophical in their approach. The philosopher who influenced the EU's chief architect Robert Schuman in particular was Jacques Maritain, with his concept of forgiveness, and his idea of personalism, of the dignity of man not being linked to race. Forgiveness— after the bloodiest war in history—was the key to it, the essence of the argument. And it proceeded as an attack on nationalism, which had led to the wars, on national sovereignty, through a pincer movement of creating the supra-national entity above and the 1931 doctrine of subsidiarity below. Then there was Erhard's social market approach, aimed at solidarity and an egalitarian society—moving money from the rich regions to the poorer regions, once again undermining national sovereignty. All this was far from being just a pragmatic response, concerned merely with coal and steel."[14]

---

[14] Peter Sutherland, conversation with the author, 1 December 2008.

The founding fathers, Schuman, Monnet, Adenauer and de Gasperi, above all, were concerned with how to ensure that there would be no repetition of the appallingly bloody conflicts which had marred European history, culminating in the apocalyptic carnage of World War One and World War Two. The establishment of the European Coal and Steel Community signed in 1951 was merely the first and most practical step towards a much more ambitious kind of Community and eventually Union.

Europe's founding values, then, were philosophical and ethical as much as economic. They included ideas about the dignity and brotherhood of humankind, about social justice, equality and solidarity, which are expressed in the Charter of Fundamental Rights. Some of these ideas are eloquently expressed in Jacques Maritain's address 'The Democratic Charter': "a community of free men is only conceivable if it recognises that truth is the expression of what *is*, and right the expression of what is *just*, and not of what is expedient at a given time for the interest of the human group…the human person is endowed with a dignity which the very good of the community presupposes and must, for its own sake, respect…the basic equality of men makes prejudices of race, class or caste, and racial discrimination, offences against human nature and the dignity of the person as well as a deep-seated threat to peace."[15]

How does all this relate to Earth Observation and the establishment of Europe's own system of satellite and in situ-based environmental monitoring, originally named GMES and now known as Copernicus?

Copernicus can be seen and understood in many different ways: it is sometimes (and erroneously in the view the present writer) envisaged as something merely technical. Of course, the monitoring system represents an immense collective technical achievement, incorporating remarkable feats of engineering. The ability to launch satellites accurately into orbit; the complexity and sensitivity of the remote sensing devices and capabilities; the information technology required to process the data: all these are extraordinary human feats and accomplishments.

We have perhaps become somewhat blasé about such things. The story, for instance of Envisat, at 8 tons and with 10 instruments on board the largest and most complex Earth Observation satellite ever launched, which remained operational from its launch in 2002 until April 2012, is little known to most European citizens but surely deserves to be celebrated.

But the significance and purpose of Copernicus obviously go beyond the technical. Its purpose is not merely to enable a certain set of technical and industrial processes to be carried forward.

---

[15] Jacques Maritain, *The Social and Political Philosophy of Jacques Maritain* (London: Geoffrey Bles, 1956) p. 157.

Copernicus can also be envisaged as a primarily economic instrument. Its purpose would then be to generate economic benefits including jobs. No doubt this is part of its purpose but does not go to the heart of the matter.

Even including the word social, as in socio-economic benefits, although necessary, is not sufficient. Copernicus promises a very wide range of social as well as economic benefits, from traffic management to the co-ordination of response to disasters. The ever more accurate monitoring of environmental changes may be able to avert or mitigate disastrous impacts, thus saving thousands or millions of lives.

Even here, however, we are not at the limits of the significance and purpose of Copernicus. We could say that we have a duty of care, to ourselves, to our immediate societies, to the wider global community, but also to the Earth itself, to the other species which inhabit it, and whose interrelations we are only just beginning to understand, and to future generations. This is the premise of an environmental ethics and it would seem that Copernicus is part of the implementation of such an ethics.

International Co-operation: GEO and GEOSS

The idea of the founding fathers was always that Europe's responsibility should go beyond Europe, and especially extend to the less developed world. Earth Observation is obviously by its nature and logic global, rather than national or regional. Access to data is patchy and often more difficult in less developed economies, where it might be most needed and where its benefits might be most appreciated. Co-ordination, co-operation and the avoidance of duplication are all essential.

For all these reasons a global system of Earth Observation has always been a desideratum. The main vehicle for it has been the Group on Earth Observations (GEO), an intergovernmental group founded in 2005 and designed to lead a global effort to build a Global Earth Observation System of Systems (GEOSS). Its origins lie in the World Summit on Sustainable Development held in Johannesburg in 2002, which "highlighted the urgent need for co-ordinated observations relating to the state of the Earth." Earth Observation, or understanding the Earth system, was seen as a crucial tool in implementing sustainable development and delivering at least some of the Millennium Development Goals set out at the UN Millennium Conference in 2000. It has special relevance to the first, eradicating "extreme poverty and hunger" and the seventh, ensuring "environmental sustainability".

A year later, at the first Earth Observation Summit in Washington, a "Declaration was adopted stating the political commitment to move towards development of a comprehensive, coordinated and sustained Earth Observation system of systems." The GEO was finally established at the third Earth Observation Summit two years later.

The GEO is a high-minded endeavour dedicated to "the well-being of humankind as a whole." For some, however, it has yet to deliver on its great promises. Roy Gibson has observed this process from its beginnings, and tells a story of disappointment and lack of political will. "The GEO started with high-powered people. I persuaded the DG of the WMO to host it. Unfortunately the political attitude was 'this is going to cost a lot of money so let's scale back.' It got drowned by committees. In fact a global system of EO satellites—a GEOSS—would cost peanuts. With EO more satellites are not the answer. Political will is the answer."[16] Perhaps the GEO is still waiting for its political champion.

Conclusion

Earth Observation developed almost by chance, or by a series of chances. The prime motive of those who embarked in balloons was not necessarily to observe the earth. Likewise instruments were attached to rockets as an afterthought. The famous Earthrise photograph, defining the fragile beauty of planet Earth and kickstarting a global wave of environmental awareness, paradoxically became the defining image of the Apollo missions whose purpose was lunar exploration.

Likewise the more limited field of meteorology provided a model and platforms for the much more ambitious project of comprehensive global Earth Observation. One of the problems with EO, as opposed to meteorological observation, is that according to a "market fundamentalist" view it is seen as a "market failure". But here the problem might be not so much market failure as an overly narrow tendency to see everything in market terms.

Viewed in a broader cultural, ethical perspective, EO can be regarded as part of our duty of care to the Earth, to future generations, and to non-human species. Perhaps the medical metaphor, notably employed when Envisat was launched "as a way of monitoring the health of the earth", is more relevant. Earth Observation is an essential tool for monitoring the health of the Earth, of planetary systems, of vulnerable features such as glaciers, polar ice-caps and rainforests, of biodiversity, of all sorts of changes to the Earth's atmosphere, oceans and land-masses, many of which are connected with human impacts in the age of the Anthropocene.

---

[16] Roy Gibson, conversation with the author, 30 May 2013.

9

The Resistances

The development of Earth Observation, and of Europe's Copernicus programme, are classic instances of the progress, and progressive use of, Western technology and science. They could even be seen as a culmination of the whole Enlightenment project, announced in Pope's famous encomium on Newton. Ever expanding knowledge, gained through mathematical, scientific and engineering prowess, is being used for the good of society as a whole.

On the whole, science and technology continue to be the dominant paradigms of our society—the most powerful guarantors of reason, truth and progress. But as we noted earlier, in the case of the mounting evidence for human-induced global warming, and wider environmental degradation, scientific and technological reason suddenly seem to have encountered limits, or resistances so strong that they threaten the entire project.

The purpose of this chapter is to examine those resistances in more detail, to see what different factors lie behind them, to inquire whether Earth Observation and the Copernicus programme in particular might hold some answers in the attempt, on the part of those who hold them to be irrational, dangerous and inspired by certain vested interests harmful to the good of society as a whole, to overcome them.

© Springer International Publishing Switzerland 2017
H. Eyres, *Seeing Our Planet Whole: A Cultural and Ethical View of Earth Observation*, DOI 10.1007/978-3-319-40603-9_9

The "Merchants of Doubt" Thesis

According to one thesis,[1] the resistance to the evidence of mounting human-induced global warming follows a pattern. This pattern was established by a series of other cases where growing scientific evidence of an environmental or health hazard was strenuously challenged by particular interests operating in a variety of overt or covert ways, often employing the services of a particular congeries of scientists, with a background in the Cold War fight against communism, predisposed to be militant against any form of government regulation.

The main cases examined by Oreskes and Conway are the link between both primary and secondary tobacco-smoking and lung cancer, the debate over acid rain and its noxious effects including the acidification of lakes and the depletion of fish stocks, and its linkage to the sulphur emissions from coal-burning power stations, the damaging impact of CFCs from refrigerators and aerosols on the ozone layer, and "Climate denialism".

In all these cases industry or government groups set about weakening or even attempting to demolish carefully assembled and peer-reviewed bodies of science pointing to serious dangers to the environment or human health. Crucial to their endeavor was the creation or employment of pseudo-independent think-tanks and the conscription of scientists who allowed political conviction—an apparently absolute belief in individual freedom and intolerance of regulation—to trump scientific scruple.

"Big Tobacco"'s Attempt to Undermine the link Between Tobacco-Smoking and Cancer

The case of the tobacco industry's attempt to discredit the steadily growing and eventually unassailable body of scientific evidence linking tobacco smoking with lung cancer is perhaps especially shocking. This link had already been proved by German scientists in the 1930s,[2] but their work was tainted by Nazi associations. In 1953 researchers at the Sloan-Kettering Institute in New York City showed that "cigarette tar painted on the skin of mice caused fatal cancers".[3] This induced consternation among the leaders of the tobacco industry. Instead of attempting to ascertain whether this important finding had relevance to humans, and thus "doing due diligence" on the safety of their product, the tobacco chiefs took a different line.

---

[1] See Naomi Oreskes and Erik M. Conway, *Merchants of Doubt: How a Handful of Scientists Obscured the Truth on Issues from Tobacco Smoke to Global Warming* (New York, N.Y.; London: Bloomsbury Press, 2010).

[2] Ibid, p. 15.

[3] Ibid, p. 14.

At a meeting on 15 December 1953, the Presidents of the four largest American tobacco companies, American Tobacco, Benson and Hedges, Philip Morris and US Tobacco agreed "to work together to convince the public that there was 'no sound scientific basis for the charges'". They would launch an Orwellian-sounding Tobacco Industry Committee for Public Information. Instead of supplying what might reasonably be considered as unbiased public information, for the public good, its aim would be "to supply a 'positive' and 'entirely pro-cigarette' message". Or as the U.S. Department of Justice would later describe it, they would attempt 'to deceive the American public about the health effects of smoking'.[4]

Essentially, the tobacco industry would stick to this line, with remarkable perseverance and in defiance of mounting scientific and epidemiological evidence to the contrary, for decades to come. As the science strengthened against them, the tobacco chiefs and their aiders and abettors, including the physicist and former President of the U.S. National Academy of Sciences Fred Seitz, resorted to a strategy which would be repeated in future. The essence of this strategy was to promote and disseminate doubt.

Science and Uncertainty

This was an intelligent strategy, Oreskes and Conway argue, because there is a misunderstanding in the public mind about the relationship between science and doubt (incidentally a misunderstanding well known to scientists such as Seitz, who cynically played on it for the tobacco industry's advantage, and later the distinguished physicists Bill Nierenberg and Fred Singer, who attempted to obfuscate the science around acid rain, the link between CFCs and ozone depletion, and human-induced global warming). The public associates science with certainty while science is riddled with uncertainty—questions are only answered in science for others to be asked.[5]

Tobacco and Uncertainty

In the case of the link between tobacco-smoking and cancer, there were always uncertainties and anomalies both at the level of individual cases and more broadly: why would one heavy smoker, or spouse exposed to second-hand smoke, contract cancer, and another not? Why was the rise in lung cancer

---

[4] Ibid.
[5] Ibid., p. 34.

rates among men greater than that among women? Why did cancer rates vary so greatly between countries and cities? These two latter anomalies could be explained—the former by the latency period of cancer, which is often diagnosed decades after a person begins to smoke, bearing in mind that historically women started to smoke later than men, and the latter by examining other environmental factors which can cause cancer—even if at the individual level uncertainty remains.

The point is that it was not necessary for science to explain all these anomalies in order to establish beyond reasonable doubt the link between both first- and second-hand smoking and cancer. According to Oreskes and Conway, this link was already established beyond reasonable doubt by the late 1950s.[6] It was confirmed in 1964 by the U.S. Surgeon General's major report Smoking and Health, summarizing 7000 scientific studies and the testimony of over 150 consultants: among its conclusions were the following: the principal cause of rise in lung cancer cases in the twentieth century was smoking, and smokers were between ten and twenty times more likely to contract lung cancer than non-smokers. In the face of this apparently conclusive evidence, the tobacco industry, aided by scientists such as Fred Seitz, continued to cast doubt on the link between smoking and cancer until well into the 2000s.

For all the extraordinary tenacity with which the tobacco industry fought its campaign of disinformation, it was eventually a losing one. In 2004 the U.S. Department of Justice brought its landmark case United States vs. Philip Morris et al. under the Racketeer Influenced and Corrupt Organizations Act (RICO). "The DOJ...sued on the ground that the tobacco companies had engaged in a decades-long conspiracy to (1) mislead the public about the risks of smoking, (2) mislead the public about the danger of secondhand smoke; (3) misrepresent the addictiveness of nicotine, (4) manipulate the nicotine delivery of cigarettes, (5) deceptively market cigarettes characterized as "light" or "low tar," while knowing that those cigarettes were at least as hazardous as full flavored cigarettes, (6) target the youth market; and (7) not produce safer cigarettes."[7] In August 2006 the District Judge Gladys Kessler issued a 1683-page opinion holding the tobacco companies liable for violating RICO by fraudulently covering up the health risks associated with smoking and for marketing their products to children. "As set forth in these Final Proposed Findings of Fact, substantial evidence establishes that Defendants

---

[6] Ibid, p. 17.

[7] http://publichealthlawcenter.org/topics/tobacco-control/tobacco-control-litigation/united-states-v-philip-morris-doj-lawsuit

have engaged in and executed—and continue to engage in and execute—a massive 50-year scheme to defraud the public, including consumers of cigarettes, in violation of RICO."[8]

Acid Rain and Ozone

Despite the eventual losing outcome, the delay between the US Surgeon General's report and the eventual finding against the tobacco industry might be considered something of a success. It ensured at least another thirty years of healthy profits for the industry, even if these were gained at the expense of the health or lives of many thousands or millions of people.

Remarkably similar tactics were employed, by different actors but with a shared basis in the visceral dislike of regulation, in campaigns to discredit the growing scientific evidence both for the link between acid rain and sulphur emissions from power stations, and for the damaging effect of CFCs on the ozone layer.

Steady scientific work throughout the 1960s and 1970s was devoted to the problem of acid rain, following the discovery of a dramatic increase in the acidity levels of rain falling in the Hubbard Brook Experimental Forest in the White Mountains of New Hampshire, a remote area far from urban centres. Scientists working both in the U.S. and in Scandinavia, where similar problems had been discovered, gradually homed in on the causes. By 1979 it was possible for the American scientist Gene Likens to state in Scientific American that the main cause for the sharp increase in acidity of rain and snow was the release of sulfur and nitrogen from burning of fossil fuels. The conclusion of an eight-year Norwegian study on acid rain reviewed by J.N.B. Bell in Nature in 1981 was that "it has now been established beyond doubt that the precipitation in southern Scandinavia has become more acidic as a result of long-distance air pollution".[9] But this apparently settled science, confirmed by the National Academy of Sciences review, suddenly became less settled as a result of a series of political tamperings and manipulations with their origin close to the centre of Ronald Reagan's White House.

---

[8] http://money.cnn.com/2006/08/17/news/companies/tobacco_ruling/
[9] Quoted in Naomi Oreskes and Erik M. Conway, *Merchants of Doubt: How a Handful of Scientists Obscured the Truth on Issues from Tobacco Smoke to Global Warming* (New York, N.Y.; London: Bloomsbury Press, 2010), p. 73.

The White House Office of Science and Technology Policy (OSTP) commissioned its own panel to review the evidence on acid rain under the chairmanship of Bill Nierenberg. Strong evidence suggests that changes were made to the Executive Summary without the knowledge or permission of the panelists, leading to a weakening of the original thrust that acid rain was traceable to sulphur emissions and that a big cut in sulphur emissions was required. These manipulations had their effect and there was no further legislation relating to acid rain during the remaining years of Reagan's presidency.

The story of the discovery of the harmful effects of CFCs from areosols and refrigerators on the ozone layer, and the successful concerted international attempt to address the problem resulting in the 1987 Montreal Protocol, revised and strengthened in 1990, is usually presented as one of the greatest successes in international pollution control. It is often contrasted with the failure to tackle rising CO_2 emissions.

There is truth in this version of events but it is not the whole truth. Once again there was an attempt to put in doubt a growing body of scientific evidence pointing to a grave environmental problem whose solution would involve regulation. In 1970 the maverick British scientist James Lovelock "documented the widespread presence of chlorofluorocarbons in the Earth's troposphere".[10] A paper by F. Sherwood Rowland and Mario Molina published in Nature in 1974 argued that the decomposition of these chemicals "would release large quantities of chlorine monoxide into the stratosphere",[11] with potentially devastating effects on the ozone layer which among other things acts as a protective shield against damaging solar radiation.

In an echo of the tobacco story, a campaign to repudiate these findings and arguments was set in train by the aerosol industry. The industry tried to argue, first, that CFCs had not entered the stratosphere, secondly that they would not break down to release chlorine, and then thirdly that the resulting chlorine monoxide would not harm ozone. At one point an attempt was made to pin the blame for ozone depletion on volcanoes.

All these counterclaims were later disproved, but the effect was to delay action. Even after the signing of the Montreal Protocol in September 1987, which mandated reductions of 50% by CFC-producing nations, there were continuing attempts to challenge the science from a congeries of anti-regulation, "market fundamentalist" think-tanks including the American Heritage Institute, the Heritage Foundation and the Marshall Institute, with the notable involvement

---

[10] Naomi Oreskes and Erik M. Conway, *Merchants of Doubt: How a Handful of Scientists Obscured the Truth on Issues from Tobacco Smoke to Global Warming* (New York, N.Y.; London: Bloomsbury Press, 2010) p. 112.

[11] Ibid.

of the physicist Fred Singer. Singer continued to argue, long after the signing of the Montreal Protocol, and even after the signing of the amended version in 1992, that there was no proof that CFCs were responsible for ozone depletion. This was a position held by virtually nobody in the scientific community, but it had traction on elements of the Republican Party in the U.S.

Global Warming

By the 1990s, the battle against regulation of CFC-containing aerosols and refrigerators was not just a losing one but an increasingly minor side-show compared to a much larger campaign: this was the beginning of the "climate wars" or the denial of global warming.

Once again we have a picture of a steadily growing and strengthening body of scientific argument pointing to a grave environmental threat, against which a campaign of obfuscation and denial was launched. As we have seen in previous chapters, the awareness of a possible impact on climate from increasing CO2 levels in the atmosphere was hardly new: it has been proposed in the 1820s by Joseph Fourier, confirmed by John Tyndall in the 1850s and 1860s and calculated in detail by Svante Arrhenius in the 1890s.

Arrhenius happened to believe that the impact of warming would be largely positive but concern over the risk of climate change began to be voiced in the 1950s and 1960s. This concern reached the heart of the U.S. Administration as early as 1965, when Lyndon Johnson gave a Special Message to Congress on Conservation and Natural Beauty which included these words: "This generation has altered the composition of the atmosphere on a global scale through… a steady increase in carbon dioxide from the burning of fossil fuels."[12]

Any push towards action on CO2 was stymied by a lack of political will in the White House, and by the kind of manipulation or spinning of a scientific report—in this case the *Changing Climate: Report of the Carbon Dioxide Assessment Committee* chaired by Bill Nierenberg—which we saw earlier in the case of acid rain.

Despite this governmental inaction, the year 1988 brought concerns over global warming to the forefront of national (in the U.S.) and international attention. This was in part because of the severe heatwave and drought in the U.S. which formed the backdrop to the climate scientist James Hansen's testimony to the Senate Committee on Energy and Natural Resources which included the statement that "the global warming is now large enough that we

---

[12] http://www.presidency.ucsb.edu/ws/?pid=27285

can ascribe with a high degree of confidence a cause and effect relationship to the greenhouse effect."[13]

1988 was also the year of the creation of the Intergovernmental Panel on Climate Change, which issued its first report in 1990. This was the point at which a concerted campaign to attack the science began, which has proved remarkably effective, at least at the level of introducing doubt into public discussion of increasingly settled science.

The increasingly settled nature of the science on human-induced global warming is revealed by the sequence of IPCC reports, issued in 1990, 1995, 2001, 2007 and 2013, showing progressively greater confidence in attributing the global warming measured since the 1940s to anthropogenic causes.

Once again the attacks on the science emanate from a cluster of think-tanks, often funded directly or indirectly by vested industrial interests, with the assistance of certain ideologically motivated scientists such as Fred Singer.

The flow of money into the think-tanks from conservative foundations and fossil fuel companies has been documented by the journalists and researchers Chris Mooney, Ross Gelbspan and Bill McKibben.[14] There have also been attempts to discredit individual scientists.

Notable is the case of the Lawrence Livermore National Laboratory atmospheric scientist Ben Santer. As the lead author of Chapter 8 of the 1995 IPCC report Climate Change 1995: The Science of Climate Change, entitled "Detection of Climate Change and Attribution of Causes", Santer found himself under attack for allegedly tampering with the report "to deceive policy makers and the public", in the words of a Wall Street Journal op-ed piece published on June 12, 1996. Santer, a scientist of unimpeachable integrity, rebutted the charges, arguing that he had made the changes in response to peer review, as he was required to do. He was supported by the chair of IPCC and many other distinguished scientists and scientific bodies. It transpired that the accusations against him had been made without any attempt to corroborate the facts.

The Concept of Journalistic Balance

This slippery concept has proved useful to climate sceptics and denialists. Broadcasters and print publications keen to maintain a reputation for impartiality like to invoke the idea of balance. In practice this is difficult to define. Does balance mean offsetting a mainstream view with an alternative or fringe

---

[13] Quoted in Naomi Oreskes and Erik M. Conway, *Merchants of Doubt: How a Handful of Scientists Obscured the Truth on Issues from Tobacco Smoke to Global Warming* (New York, N.Y.; London: Bloomsbury Press, 2010) p. 184.
[14] Ibid, p. 246.

one, even where that alternative or fringe view has vastly fewer adherents? Even the greatly respected broadcaster Ed Murrow was, arguably, misled by the concept of balance when he gave credence to the tobacco industry's attempts to deny the links between smoking and lung cancer. Some speculated that Murrow may have been influenced by his own smoking habit. Much more recently the widely respected BBC has abused "balance" by apparently equating the views on climate change of mainstream science and maverick denialists; the 2014 Executive Report on Science Impartiality commissioned by the BBC Trust and chaired by the geneticist Professor Steve Jones criticised the BBC's coverage of climate change for giving "undue attention to marginal opinion".

Cultural Resistance

However persuasive the Merchants of Doubt thesis, this strategy could only have traction given certain cultural and psychological preconditions.

First we can examine the cultural preconditions. As we have seen, the dominant paradigm since the seventeenth century has been of the earth as an insensate, mechanical resource suitable for infinite exploitation by humans. Of course, even in earlier times, some voices were raised in concern against this view, and more recently it has come under much more concerted challenge from a variety of sources, many grouped under the general heading of "environmentalism". Examples of this are the rise of green political parties, the formation of environmental NGOs with wide reach, and international bodies operating under the aegis of science such as the IPCC.

Environmentalism may have become part of the mainstream, but it is still probably not the most powerful strand in contemporary thinking. The low profile of environmental concerns, and the limited success of Green parties, in the majority of elections in developed countries seems to bear out this view.

Julian Simon and Cornucopianism

Here it is worth going back to examine in more detail the environmental debates of the 1980s. The warnings of figures such as Paul Ehrlich and groups such as the Club of Rome (in its Limits to Growth report) about environmental degradation and the depletion of natural resources were met by a robust resistance in what became known as "Cornucopianism".

This school of thought associated with such writers as Julian Simon and Wilfred Beckerman holds that there are very few "limits to growth". Essentially, the earth's resources combined with ever-increasing technological ingenuity are more than sufficient to cope with increasing human population and demands. Simon addresses contemporary environmental concerns around such issues as population, food production, energy, pollution, natural resources and so on. In all these cases he concludes that fears are misplaced: "there is compelling reason to believe that human nutrition will continue to improve into the indefinite future, even with continued population growth[;]...past experience [gives no] reason to expect natural resources to become more scarce[;]... the long-run future of energy is at least as bright as that of other natural resources."[15] One of Simon's chapters is provocatively titled "Will we run out of oil? Never!". He is a strong advocate of nuclear power.

It is worth pointing out that some of Simon's robustly optimistic assertions in The Ultimate Resource 2 have already been proved wrong: on ozone depletion Simon's view is "The best principle might be: 'Don't do something. Stand there.'...The long-run data on ozone show no trends that square with public scares."[16] These words which appear in the 1996 edition of The Ultimate Resource 2 seem the opposite of prescient.

Simon is now in his 80s; his work is being carried forward by figures such as the British popular science writer Matt Ridley and the Danish statistician Bjorn Lomborg. One of Ridley's themes is the apparent contradiction between the facts of human progress—a vastly increased population living far better than in past ages—and the continuing belief in impending doom. Ridley is not a global warming denier but believes the potential negative impacts of global warming have been exaggerated.[17]

The Cornucopians have won some arguments: in particular, Paul Ehrlich's prediction in early editions of The Population Bomb that in the 1970s hundreds of millions of people would starve to death has not come to pass. Despite theories of "peak oil" supplies of fossil fuels have recently been increasing, especially with the discovery of shale gas deposits in north America.

The robust assertions of the human capacity to cope with almost any environmental constraint, which are close to the heart of Cornucopianism, are not only apparently reasonable, in stressing the extraordinary adaptive power of humans and human technology, but also reassuring. The denial of limits fits in with what the social scientist John Dryzek calls the Promethean strain in environmental thinking. The human project is one long success story of adaptation to environmental conditions and there is no reason to suspect this will ever end. Simon even imagines humans in the far distant future being

---

[15] Julian Simon, The Ultimate Resource 2 (Princeton N.J.: Princeton University Press, 1996), pp. 5–6.
[16] Ibid.
[17] See Matt Ridley, The Rational Optimist (London: Fourth Estate, 2010) pp. 328–41.

able to compensate for the lost energy of the dying sun.[18] Optimism does not get much more sunny than that. Of course, one might also note that this is precisely the kind of optimism mocked so mercilessly in Voltaire's Candide, in which the relentlessly optimistic Dr Pangloss is able to put a positive spin on any turn of events, including his own infection with syphilis (on the whole, imports from the New World have been beneficial).

Cornucopianism or Prometheanism or techno-optimism—the world view that the earth's resources combined with human technological ingenuity are basically inexhaustible and environmental problems greatly exaggerated—is arguably the dominant world-view of our age, comparable to that of the Catholic Church in the time of Galileo. In the same way that the Catholic geocentric view was buttressed by the immense influence, power and wealth of the Catholic Church, including the murderous Inquisition, the Cornucopian view is supported by many vested interests close to the heart of twenty-first century capitalism, most notably the fossil fuel lobby. The political powers-that-be have to some extent aligned themselves against these vested interests by signing instruments such as the Kyoto Protocol and making commitments to reduce fossil fuel use, most recently at the Paris COP21 Conference, but so far these commitments have not been backed up by urgent action; a partial exception might be made for the bilateral agreement reached by Presidents Obama and Xi Jinping in November 2014 to commit to targets for cuts in their nations' carbon emissions.

Psychological Resistances

Now we need to move from cultural to psychological resistances. It is not just that the Cornucopian view, which encourages business as usual in the exploitation and despoliation of the earth's resources, seems right, rationally, to many people but also that it feels right, psychologically. We do not just believe it but we want or need to believe it, because the alternative is so uncomfortable. The discomfort is complex; if the non-Cornucopian view is correct, that means not just that we may have to change our behaviour and habits, in inconvenient ways (not so much use of automobiles and aeroplanes, or even a more general move away from consumerism), but also that we might have to face up to moral or ethical shortcomings: we have been living irresponsibly, in relation to other people in our own world suffering more acutely

---

[18] Julian Simon, *The Ultimate Resource 2* (Princeton N.J.: Princeton University Press, 1996), p. 181.

from environmental externalities, in relation to our children and children's children, who may inherit a world irreparably damaged by our actions, and in relation to other species and the entire natural world which we are systematically degrading.

Denial

Here the psychological concept of denial can be of assistance. Denial is one form of the defence mechanisms (originally identified by Sigmund Freud) by which the self or the ego wards off uncomfortable or unbearable thoughts, perceptions or truths. For example, one of Freud's early patients found it unbearable to acknowledge that she was in love with her brother-in-law. Instead she suffered from "hysterical" symptoms. In the essay The Loss of Reality in Neurosis and Psychosis Freud writes that in both neurosis and psychosis "there is no lack of attempts to replace a disagreeable reality with one that is more in keeping with the subject's wishes."[19] The concept of denial was further developed by later psychiatric and psychoanalytical writers such as Elisabeth Kübler-Ross. Kübler-Ross identified five stages of "coping mechanisms at the time of a terminal illness"[20] of which the first is denial and isolation.

The Australian scholar Clive Hamilton has attempted to identify different types of denial which are in operation around the question of climate change. The first is simply distraction: people find it easier not to pay attention to troubling reports, by ignoring stories in newspapers or switching off the TV or radio when it touches on such themes. A second kind of denial, according to Hamilton, is "deproblematising", which occurs when the threat of climate change is recognised, but then devalued as being either exaggerated or easily soluble, as other such threats have been in the past. Hamilton refers to a third strategy of denial as "distancing": this means "emphasising the time lag before we feel the consequences of warming".[21] He calls this a form of "wishful thinking" because it implies the thought that the problem will cease to exist or be solved merely because it is so far off.

Another kind of denial is represented by the popular British TV presenter Jeremy Clarkson. In his long-running BBC TV motoring series Top Gear Clarkson took delight in mocking environmentalists, deriding public trans-

---

[19] Sigmund Freud, *Complete Psychological Works* ed. James Strachey (London: Hogarth Press: Institute of Psychoanalysis, 1953–74), Vol.19 p. 187.

[20] Elisabeth Kübler-Ross, *On Death and Dying* (London: Tavistock Publications, 1973), p. 33.

[21] Clive Hamilton, *Requiem for a Species: Why We Resist the Truth about Climate Change* (London: Earthscan, 2010), p. 122.

port and even "promising to run down cyclists".[22] Clarkson seemed to speak for the teenage boy, or possibly even toddler, trapped inside each (predominantly male) English breast—the part of the self devoted to self-gratification in defiance of all authorities.

Yet another form of denial, Hamilton suggests, is blame-shifting; this means the problem is recognised but considered to be somebody else's responsibility. Examples are a tendency among smaller countries to blame larger ones and a general proclivity to demonise China with reports of the world's most populous country "building a new coal-fired power station every week."

It is a moot point whether cultivating optimism—often seen as a desirable or even compulsory social norm (as in the US where "have a nice day" often sounds, to Europeans, more like a command than a greeting, or as in the case of the attempted rebranding of the 2009 Copenhagen Climate Conference as "Hopenhagen")—counts as a form of denial. Certain forms of forced or willed optimism may be helpful in getting things done or avoiding paralysing despair. A delusional optimism—adhered to in defiance of facts—can only lead to distancing from reality. In Freudian terms this would count as a form of psychosis.

Science and Advocacy

A related question concerns the responsibility of scientists to speak out, or remain neutral, on the implications of climate science. A body of opinion within the scientific community holds that it is the responsibility of scientists purely to get on with doing the science. Any confusion of this with advocacy or political involvement risks contaminating the purity and efficacy of the science. An articulate exponent of this view is the British climate scientist Tamsin Edwards. Edwards believes that "advocacy by climate scientists has damaged trust in the science". She says she "cares more about restoring trust in science than calling people to action".[23]

On the other side of argument stand scientists such as James Hansen, longtime head of NASA's Goddard Institute of Space Studies in Manhattan, and Michael Mann, Director of the Earth System Science Center at Pennsylvania State University. They have for many years rejected the idea that scientists should maintain an impartial position and avoid value-laden language or political posturing.

---

[22] Ibid, p. 124.
[23] http://www.theguardian.com/science/political-science/2013/jul/31/climate-scientists-policies

Hansen, who now runs the Programme on Climate Science, Awareness and Solutions at Columbia University, has backed up his belief that science has a responsibility to communicate the gravity of the climate crisis to citizens by engaging in acts of civil disobedience; he has twice been arrested following protests in front of the White House against the extension of the Keystone XL pipeline.

Michael Mann is another who believes, he wrote in the New York Times in January 2014, that "it is no longer acceptable for scientists to remain on the sidelines." A study he co-wrote in 1998 showed that Northern Hemisphere average warmth had no precedent for at least 1000 years; this in his own words led to him being "hounded by elected officials" and "threatened with violence". He makes a provocative parallel with the advice from the US Department of Homeland Security to citizens to report anything dangerous that they witness: "if you see something, say something." Scientists, the overwhelming majority of them, can see a serious, ever-deepening threat: it is their responsibility, according to Hansen, Mann and others, to speak out.

Science and Rhetoric

One problem underappreciated by the proponents of scientific neutrality such as Tamsin Edwards is that if scientists abstain from advocacy, the rhetorical space, as it were, will be quickly filled by other voices with fewer scruples about purity of method. This can be seen clearly in the field of the media. Scientists unskilled in the arts of communication (that is in communicating one particular point of view) are often pitted against highly skilled advocates from the denialist or "sceptic" camp. The impression given to the public will often be that the scientists are cautious and careful to hedge around every assertion with numerous qualifications while the sceptics or denialists show no such misgiving.

Here we can remember Edmund Burke's famous call to action, in his Thoughts on the Cause of the Present Discontents (1770): "When bad men combine, the good must associate; else they will fall one by one, an unpitied sacrifice in a contemptible struggle."[24]

---

[24] http://www.econlib.org/library/LFBooks/Burke/brkSWv1c1.html

Reversal of the Burden of Proof

The well-orchestrated denialist or sceptic campaign conducted against mainstream climate science and the work of the IPCC in particular has tended to put the mainstream climate science community on the back foot. Strangely the burden of proof appears to have been reversed: it is no longer the denialists who called upon to prove their largely tendentious claims, but rather the consensus position, supported by the overwhelming majority of climate scientists, which seems to be in the dock.

This may be related to an older debate about science, or philosophy, and rhetoric, which goes back at least as far as Plato. In the Republic, Plato's protagonist Socrates goes so far as to banish the poets from the ideal state, on the grounds first that their productions are far removed from truth and secondly that they employ emotional manipulation.

Should Copernicus Be Neutral?

A key question for decision-makers in the EC and in ESA is whether Copernicus should be neutral. The Director General of ESA, Johann-Dietrich Wörner, has declared his view that Copernicus should be non-neutral as well as being non-political.[25] A parallel can be made with health: a health system, or health monitoring system, can be non-neutral, that is to say firmly on the side of health, at the same time as being non-political.

How EO Can Help

Satellite Earth Observation provides a steady and comprehensive flow of data covering all aspects of the environment, terrestrial, marine, atmospheric. EO does not substitute for in situ measurement, using buoys, balloons etc.; both are essential. The advantage of satellite observation is the ability to monitor and measure parts of the environment that are beyond the reach of in situ sensors: the oceans, the ice caps, glaciers, rainforests, the upper atmosphere.

Earth Observation is not a matter of theory but of empirical measurement consolidated over time. This is also an advantage given the highly politicised nature of the climate debate. The data from Earth Observation are not in

---

[25] Johann-Dietrich Wörner, conversation with the author 11 March 2015.

themselves contestable, even if models and theories derived from them may be. It is interesting that many of the leading "climate denialists" have been physicists (or in one case a statistician), and that very few have been climate scientists. Physics can be seen as the most "macho" of the sciences and the most theoretically powerful. It deals with the world at a certain level of abstraction and detachment. Earth Observation by contrast deals only with the facts on the ground, or the sea or the air. It is a patient, slow, humble process.

So Earth Observation may offer a helpful way of depoliticising the climate, and wider environmental, debate. While theories or models may provide different views or predictions on say the rate of sea level rise, or the melting of the icecaps and glaciers, or concentrations of greenhouse gases in the atmosphere, or rates of deforestation in Amazonia, Earth Observation simply measures these things. This is not to say that measurements themselves cannot be sometimes in need of correction; the methods and methodologies of data collection need themselves to be constantly monitored. But that is part of the method and process of science, rather than politics. One view would be that Earth Observation can be of assistance in establishing limits within which the political debate can be conducted; certain kinds of political arguments or positions can be simply disqualified as being outside these firmly established parameters.

10

The Aesthetic Potential
of Earth Observation

Earth Observation has rarely been considered from the point of view of aesthetics. The focus has been firmly scientific and objective. A magnificent achievement of the hard scientific disciplines of physics, engineering, mechanics, EO satellites have been regarded as techno-scientific tools, designed to generate quantifiable data which can be processed and recombined, for utilitarian ends. This will undoubtedly remain the dominant paradigm under which EO is considered, but it is not the only one.

Earth Observation and Remote Sensing

One reason for the neglect of the aesthetic perspective may lie simply in the name, or the change of nomenclature from the old-fashioned "remote sensing" to the current "Earth Observation". Perhaps to the scientific or technical mind there is little difference between the two, as the emphasis may be on the data produced, represented as digital information. But to a more poetically or philosophically informed sensibility there is a considerable difference. "Sensing" linguistically speaking is cognate with sensitivity and even sensibility. To sense something is to feel and register it, with all the senses in play, especially perhaps those of smell and touch. It suggests intuition as much as cognition. The usually dominant sense of sight is downplayed.

All this of course is to speak from a human perspective. But much of the remote sensing done by sensors is not primarily visual. Gases in the atmosphere are sensed in a manner more analogous with smell than sight.

© Springer International Publishing Switzerland 2017
H. Eyres, *Seeing Our Planet Whole: A Cultural and Ethical View of Earth Observation*, DOI 10.1007/978-3-319-40603-9_10

Radar interferometry operates through the combination of different sets of radar waves which are reflected back from objects. The effect is similar to hearing an echo.

Observation has very different connotations. It is of course primarily visual, but beyond that connotes surveillance. To be "under observation" suggests either a medical condition or, perhaps, being suspected of wrong-doing. Surveillance, connected to observation, might originally have been neutral but now has many negative connotations. Recent revelations as a result of leaks from the whistle-blower Edward Snowden have revealed a far more extensive kind of surveillance being carried on in western democracies than many previously suspected. The mass surveillance of citizens' electronic communications and phone conversations brings us uncomfortably close to the world imagined by George Orwell in 1984, or to the world which actually existed in Communist East Germany.

An Aesthetic History of Lunar and Earth Observation

Many EO images, to take the most obvious example, take the form of photographs, which can be viewed aesthetically as well as used instrumentally. Exhibitions of such photographs, usually coloured for aesthetic effect, can sometimes be found in the public areas of space institutions such as ESA or ESPI but are rarely shown beyond such confines. Beyond that, it is worth looking at EO from the viewpoint of the artist, poet or philosopher, not just the engineer or scientist.

We can say that space in general has always possessed this potential for an aesthetic view. Early examples include Lucian's fanciful account of a trip to the moon, a few rough sketches by Leonardo da Vinci (c. 1500) the beautiful and more detailed wash drawings by Galileo (c. 1610) and the roughly contemporaneous (in fact slightly earlier) drawings by Thomas Harriot, three engravings by Claude Mellan based on sketches by the astronomer Pierre Gassendi and the beautiful and painstakingly accurate large pastel drawing The Face of the Moon by the English portraitist John Russell (1793–1797) which hangs in Soho House in Birmingham.

Russell was not just a fashionable portrait painter but also a passionate astronomer, as he explained in a letter of February 19, 1789, to Dr Thomas Hornby, Observer of the Radcliffe Observatory at Oxford: "About twenty-five years since, I first saw the Moon through a telescope, which I now recollect must have been about two Days after the first quarter; you will conclude how much struck a young Man conversant with Light, and Shade, must be with the Moon in this state… a few Days after I made a small Drawing, but the Moon being at the Full, I was not struck in the same manner, and I made no more attempts, till an accidental possession of a powerful Glass awakened my attention to this beautiful Object once more, and for several years I have lost few opportunities when the Atmosphere has exhibited the Object of my study and imitation."[1] His image is striking for its scientific accuracy, showing a particular (waxing, gibbous) phase of the moon and depicting its features such as craters and "lakes" in meticulous detail. At the same time Russell is also exercising his artistic sensibility, as well as great skill: he says that the poor quality of existing maps of the moon "led me to conclude I could produce a drawing in some measure corresponding to the feelings I had upon the first sight of the gibbous moon through a telescope."[2]

In this sense Russell's accurate pastel drawing is much more modern than the largely subjective poems about the moon which abound from the renaissance onwards, for example the famous sonnets by Sidney, Daniel and Keats.

Different Art-Forms and Media

Clearly the most obvious art-form in this connection is photography. But both EO images and, at a more conceptual level, the idea of observing the earth from space, with all its ambiguous potential, for environmental protection as well as surveillance, could be inspiring to visual artists, poets, novelists, philosophers, musicians. It is perhaps surprising that EO has so far generated relatively little in the way of aesthetic and artistic response.

Earthrise

The most famous EO image is of course the 1968 Earthrise photograph, which has become a kind of cliché of environmental discourse. However apparently hackneyed, though, the Earthrise photograph remains an inescapable point of departure for this kind of discussion. It has the characteristics of a potent aesthetic image in generating a huge range of possible interpretations. Aesthetic images are not a means to an end; they are in a sense an end in themselves, and inexhaustible, as Yeats put it in his great poem 'Byzantium'—"those images that yet/fresh images beget, that dolphin-torn, that gong-tormented sea."

[1] http://www.andrewgrahamdixon.com/archive/readArticle/71
[2] http://www.independent.co.uk/arts-entertainment/art/great-works/russell-john-the-face-of-the-moon-17937-744427.html

Earthrise, a single photograph taken during the fourth orbit of the moon by the Apollo 8 mission on the morning of Christmas Eve 1968, is an ambiguous image. Some of the ambiguity may arise from the circumstances in which the photograph was taken. It is in fact noteworthy that this photograph was "taken", by a human photographer or photographers (the astronauts Frank Borman, Jim Lovell and Bill Anders), rather than automatically generated. In an interview with Jeffrey Kluger of Time magazine on 24 December 2013 (the 45th anniversary of the taking of the Earthrise photograph), Jim Lovell spoke as follows: "My opinion at the time was that it would be a great picture...But I didn't comprehend that, in today's language, it would go viral— that it would be the capstone and message of the mission."

Joseph Cornell's Collages: Observations of a Satellite (1) and Weather Satellites

In 1960 and 1965 the eccentric American artist Joseph Cornell (1903–1972) created two collages with the titles 'Observations of a Satellite (1)' and 'Weather Satellites'. From an early age Cornell had been fascinated and sometimes terrified by space and astronomy. His sister Elizabeth recalled that while at home on vacation from boarding school Joseph had woken "shivering awfully, and asked to sit on my bed. He was in the grips of panic from the sense of infinitude and the vastness of space as he was becoming aware of it from studying astronomy."[3]

The two collages are very similar: central to both is the image of a humming-bird trapped in a glass bell jar, apparently looking up at the dusky sky, in which a green icosahedron is suspended. In the upper right corner of the bell jar in both collages can be made out the reflection of a cork-stoppered glass bottle. In the earlier collage, 'Observations of a Satellite (1)', the humming-bird shares the confined space of the glass dome with the image of the Infanta from Velázquez's Las Meninas, and the sky is divided by a grid-like pattern. The tiny humming-bird's ability to migrate through vast expanses of air, and look back at the earth, somehow chimes with the soaring of the satellites, miniature objects when set against the vastness of space. The icosahedron may refer to Kepler's Harmonices Mundi, in which the great German astronomer, admired by Cornell, analysed such geometrical forms. Both of these exquisite collages convey a powerful sense of wonder and longing, as well as the sense of confinement within narrow bounds which was so much a part of Cornell's life.

---

[3] https://archive.org/stream/josephcornellcos1222corn/josephcornellcos1222corn_djvu.txt

The Power of Images and Poetry

The power of photographic images should not be underestimated. As we have seen, the single Earthrise photograph became one of the crucial factors in creating the "first wave of environmentalism" of the 1960s. Nick Ut's 1972 photograph of a naked Vietnamese child fleeing an American napalm attack during the Vietnam War may have done more to change US public opinion about the war than anything argued by politicians. As Susan Sontag argued in On Photography, "Something we hear about, but doubt, seems proven when we're shown a photograph of it."[4] The power of this observation was reinforced in September 2015 by the effect of the harrowing photographs of the drowned body of 3-year-old Syrian refugee Aylan Kurdi on a Turkish beach.[5] Suddenly public opinion, and public policy, in various European countries, especially the UK, appeared to shift from indifference to sympathy. The strict control over the access of journalists to warzones in conflicts after the Vietnam War is also a tacit acknowledgement of the truth of Sontag's remark.

World War One Poets

Though we are more familiar with the power of visual images, the power of poetry can be considerable. It is arguable that the public attitude towards the First World War in Britain, generally positive for most of the 1920s, was dramatically changed by the publication and gradually wider dissemination of war poems graphically describing the carnage and registering the terrible waste of life, by poets such as Wilfred Owen and Siegfried Sassoon, and the war autobiographies Undertones of War, Goodbye to All That and Memoirs of an Infantry Officer by the poets Edmund Blunden, Robert Graves and Siegfried Sassoon. Mention should also be made of the work of the war artists including John Singer Sargent, John Nash, Wyndham Lewis and others.

Though some of Sassoon's poems were actually published during or immediately after the war, in 1917 and 1919, and some of Owen's (posthumously) in 1920, their impact only became really strong in the politicised climate of the 1930s and 1940s. This was greatly aided by the publication of anthologies such as Frederick Brereton's Anthology of War Poems of 1934 and Robert Nichols' Anthology of War Poems 1914–18 (1943). A further step in the dissemination and influence of these poets came in the anti-war climate of

---

[4] http://www.macobo.com/essays/epdf/onphotography.pdf
[5] http://www.theguardian.com/commentisfree/2015/sep/06/photograph-refugee-crisis-aylan-kurdi

the 1960s and with the greatly resonant premiere in the rebuilt Coventry Cathedral on 30 May 1962 of Benjamin Britten's A War Requiem. This is a large-scale choral and orchestral work with three soloists setting nine poems by Owen. The solo singers for whom the work was written were the British tenor Peter Pears, the German baritone Dietrich Fischer-Dieskau and the Russian soprano Galina Vishnevskaya. Shortly before the premiere Vishnevskaya was refused permission to travel by the Soviet authorities; her place was taken by the English soprano Heather Harper.

North American and southern hemisphere premieres followed just over a year later, in July 1963. Decca issued a recording of the War Requiem in 1963, with Pears, Fischer-Dieskau and Vishnevskaya as soloists, which sold over 200,000 copies in the first five months of release, a most remarkable figure. Britten is reported to have said, in a letter to his sister, "I hope it'll make people think a bit." On the title page of the score he quoted these words by Owen:

> "My subject is War, and the pity of War.
> The Poetry is in the pity...
> All a poet can do today is warn."

It was in the 1960s also that Joan Littlewood's satirical musical drama "Oh, What a Lovely War!" opened in the East End of London and that the World War One poets including Sassoon and Owen began to be widely taught in schools. By the 1970s and 1980s, as the present writer can attest, their view of the war had become the dominant one.

This story of the war poets shows a particular artistic expression can radically alter prevailing public opinion of an event or even of history. In this particular case the pendulum has swung so far that at the time of the centenary of the beginning of World War One there were calls for a more "balanced" approach to the war which recognised its component of noble sacrifice.

Eco-Poetry, Land Art

A school of ecological poetry has arisen, led in the US by such poets as Gary Snyder, Robert Bly and William Stafford, and in the UK by Ted Hughes and Alice Oswald. The anthology *Earth Shattering: Ecopoems* edited by Neil Astley at Bloodaxe Press is part of this current. Its influence has so far been quite limited compared with that of the war poets.

More influential has been the development of 'Land Art'. This is a movement connected to the conceptual art movements of the 1960s in which landscape and artwork become so closely intermingled as to be almost indistinguishable. Perhaps the most famous example is Robert Smithson's Spiral Jetty (1970) which consists of a jetty made of earth constructed on the shores of Great Salt Lake. At times the jetty is completely covered by water. Preceding it by three years is the British artist Richard Long's A Long Line Made by Walking, a photograph of a line made in grass by the artist's repeated footsteps. Long's works have been described as "primitive expressions of man's relation to the earth."[6]

Characteristics of land art are sensitivity to the contours and particularities of the earth and a relinquishing of the attempt to make any permanent mark upon it. There is often either explicit or implicit acknowledgement of the gentler attitude to the earth shown by earlier cultures and civilisations.

Closely connected to Land Art are the Italian Arte Povera movement especially the work of Giuseppe Pennone and interventions such as Joseph Beuys' 7000 Eichen (1982) in which the German artist planted 7000 oak trees.

Pop Music

In pop music the English singer, musician and actor David Bowie (1947–2016) found repeated inspiration in space, especially in the early part of his career, including the single Space Oddity (1969) and the album The Rise and Fall of Ziggy Stardust and the Spiders from Mars (1975). But in his case it is not so much the view of the Earth from space which is an inspiration, but even more the idea of being an alien either suspended far above the planet or dropping to Earth from Mars, reflecting an inner sense of alienation.

Dramatic Potential: Phylae and the Story of Envisat

One artform that is rarely if ever invoked in connection with EO is drama. However, the stories of the launches and operating missions of satellites certainly have dramatic potential. Rather belatedly, it seems, with the comet lander Phylae, the European Space Agency discovered the anthropomorphic potential of unmanned space missions. The adventure of Phylae seemed to capture imaginations in a way rarely seen with satellites; this was partly

---

[6] http://www.tate.org.uk/context-comment/articles/richard-longs-line-made-walking

because of the consciously anthropomorphic language used to describe the lander's activities. For instance, it was said to "awake from hibernation", "get back in touch", "phone home again."

ESA's Envisat, was the largest civilian EO satellite ever launched, when it was sent into a sun-synchronous polar orbit using an Ariane 5 launcher on 1 March 2002. It carried advanced imaging radar, radar altimeter and temperature-measuring radiometer instruments which extended the data sets of the ERS mission it replaced. This was supplemented by new instruments including a medium-resolution spectrometer sensitive to both land features and ocean colour. Envisat also carried two atmospheric sensors monitoring trace gases.[7]

The Envisat mission was inherently dramatic given the enormous size and cost of the satellite (over 2 billion euros) and the complexity of its payload of instruments. For instance, if the launch of Envisat had failed, the impact on the European Space Agency might have been heavy. This impact would clearly have been emotional as well as financial and practical, as many scientists and engineers had invested considerable parts of their working lives in the mission.

In the event both the launch and operation of Envisat were a success. ESA lost contact with Envisat on 8 April 2012, more than five years after the estimated completion date of the mission.

The aims of Envisat were as follows: "to provide for continuity of the observations started with the ERS satellites, including those obtained from radar-based observations;

- to enhance the ERS mission, notably the ocean and ice mission;
- to extend the range of parameters observed to meet the need of increasing knowledge of the factors determining the environment;
- to make a significant contribution to environmental studies, notably in the area of atmospheric chemistry and ocean studies (including marine biology).

Coupled with these were two linked secondary objectives:
- to allow more effective monitoring and management of the Earth's resources;
- to better understand solid Earth processes."[8]

---

[7] https://earth.esa.int/web/guest/missions/esa-operational-eo-missions/envisat
[8] https://earth.esa.int/web/guest/missions/esa-operational-eo-missions/envisat/objectives

In all these areas Envisat made notable contributions.

In a 2001 press release, ESA revealed that "Envisat will orbit the Earth 14 times a day and probe every corner of the earth for environmental pollution and climate change. It will measure and analyse greenhouse gases in the atmosphere, locate environmental polluters, identify ocean currents and algae growth and keep a watch on the ozone." Here there is an attempt to dramatise the mission, portraying Envisat as a kind of detective of the skies. But it appears that as time went on there was less attempt to engage public attention at this more general and emotional rather than purely scientific level—the level at which a wide public regularly engages with manned space missions. The reluctance to emphasise and communicate the inherently dramatic nature of the mission could be one reason why Earth Observation languishes behind other uses of space in the popular imagination.

Emotional Connection

One great advantage—perhaps the greatest advantage—of the arts is that they operate on an emotional and even visceral level, as well as an intellectual level. Put another way, the arts can concentrate complex ideas into aesthetically and emotionally satisfying and cogent forms. The power that art has over people has been debated since the time of Plato and Aristotle. Both agreed that drama in particular had a strong emotional effect on audiences; they disagreed on whether that effect was healthy or unhealthy. Aristotle in the Poetics argued that the effect of tragedy was healthy, inducing what he called a catharsis or purging of the emotions of pity and terror which tragic plays induced in the audience. Plato as we have seen was much more suspicious of this emotional power.

Psychoanalysis and Emotional Connection

An analogy can be made here with psychoanalysis. One of Freud's early discoveries was that in order to effect psychological change, mere intellectual understanding was not enough. Change occurred through what he called the transference, the messy and tangled process by which feelings and modes of relationship are carried forward from the past into the relationship with the therapist. Without the transference, a patient might gain understanding and insight but behaviour would not change.

The point is that it is necessary to change both hearts and minds. Minds may be won over by "knowledge", facts, statistics and intellectual arguments, but hearts need emotional connection, and behaviour will not change until both hearts and minds are won over. This seems especially relevant to the climate change debate, where a strange kind of cognitive dissonance appears to prevail. A majority of people in Western countries are convinced of the reality of man-made climate change, but very few are prepared to act on this knowledge.

Conclusion: Implications for Communication

There is surely a huge untapped potential for the aesthetic use of EO images, and indeed for ways of communicating in relation to Earth Observation which are not purely fact- or science-based, but which bring into play beauty, poetry, drama and a whole range of emotional responses. Earth Observation can be a way not just of providing data about the state of the earth, but of engaging emotionally with it, as happened with the Earthrise photograph, and has been repeatedly confirmed in the accounts of astronauts' feelings on looking back at the earth from space. Several have said that the only adequate way of articulating these feelings is through poetry.

11

EO for Whom? Towards an Environmental Democracy

Competing Ideas of Science and Environment

There is clearly no single idea of Earth Observation, or of environment, or even of science. A popular view of science seems to be that it provides answers. Scientists themselves are more inclined to think in terms of testable hypotheses which constantly give rise to new questions. Beyond this, the view of science as offering a kind of absolute truth, existing in a world of pure forms beyond any contingent contamination, as dreamed of by Plato, has been challenged by the sociological critique of science, especially the so-called Edinburgh school of the Sociology of Scientific Knowledge (SSK) and a variety of postmodern thinkers who see science as at least in part a social construct.

Already Thomas Kuhn saw that science is at least to some extent socially constructed—that the direction and even the nature of scientific research is influenced by social and political factors, rather than representing the inevitable unfolding of eternal laws of nature.[1] Sheila Jasanoff has written that "science and technology operate as...*political agents*."[2] Feyerabend suggests that there is no single "scientific method" but many methods and approaches.[3]

---

[1] See Thomas Kuhn, *The Structure of Scientific Revolutions* (Chicago; London: University of Chicago Press, 1968).

[2] http://isites.harvard.edu/fs/docs/icb.topic251437.files/Fall%202007%20papers/Jasanoff_papers/Jasanoff_States_of_Knowledge_CH2.pdf

[3] See Paul Feyerabend, *Against Method: Outline of an Anarchistic Theory of Knowledge* (London: NLB, 1975).

© Springer International Publishing Switzerland 2017

H. Eyres, *Seeing Our Planet Whole: A Cultural and Ethical View of Earth Observation*, DOI 10.1007/978-3-319-40603-9_11

At the very least, it must be admitted that science is the product of a particular culture and world view. The objectifying view which Western science traditionally takes of nature, and to some extent of human beings, is quite unfamiliar and indeed unpalatable, as we have seen, to many other cultures.

"Environment" is also a term which is interpreted and understood in many different ways, even within the Western tradition. Does environment consist of natural resources and ecosystem services which exist essentially outside and in contradistinction to human beings and entirely for their benefit? Or is environment an interlinked network of species (perhaps reminiscent of the Great Chain of Being alluded to by medieval and renaissance thinkers) and habitats from which no one species can isolate itself? Is environment our beautiful mothering home presided over by a benign God, as it appears in Pope Francis's recent encyclical Laudato Si?

These differing views would naturally give rise to different attitudes. Is environment there simply to be managed for the benefit of humans, or does it have intrinsic value? Does it even have spiritual value and importance, as argued by Pope Francis?

Thus it is hardly surprising that there would be competing views of Earth Observation and the purposes it serves. In some ways satellite Earth Observation would seem to fit best with a detached, managerial view of environment. In the most dystopian of views, Earth Observation could even be used for ever more unsustainable exploitation of the environment. Individual large-sized fish and aquatic mammals can already be monitored from space; it would only be a matter of the information falling into the wrong hands for it to be used for exploitation rather than protection.

The use of Earth Observation satellite data for intensive agriculture is certainly very different from this, in that the aim is for sustainability and best and most timely use of limited resources, especially water. But this certainly operates on a managerial paradigm. We would like to go further and argue that there is also the potential for Earth Observation to be used for stewardship as much as management. The purpose of Earth Observation could be as much to monitor the health of the planet—for all species and for generations still to come as well as those living in the present—as to ensure its efficient and profitable exploitation for human benefit in the short term.

EO as Market Failure; The Limits of Market Thinking

It has become commonplace, or even the norm, to speak about environmental problems as "market failures". When Nicholas Stern wrote his 2006 Stern Review for the British government on the economics of climate change, he described climate change as the greatest and most wide-ranging market fail-ure ever seen. This way of speaking went virtually unchallenged, even among environmentalists, who generally welcomed the report. Very few people pointed out that a potentially catastrophic destabilizing of the climate on earth, impacting especially the poor and future generations, as a result of human actions, could be seen not just as a market failure but as an ethical one, a crime or an unparalleled act of hubris.

This way of speaking is obviously well entrenched—perhaps almost unchal-lengeable. Perhaps it is worth unpacking some of its underlying assumptions. The chief of these is that market mechanisms are the normal or even natural way of organizing human affairs. Everything ought to function as a market; if something does not, it constitutes a "market failure".

But not everything works or should work according to market principles, for all their many and great virtues. This is the point made by the American moral philosopher Michael Sandel in What Money Can't Buy: the Moral Limits of Markets. Sandel argues that "over the past three decades, markets—and market values—have come to govern our lives as never before…We drifted from having a market economy to being a market society."[4]

What is wrong with a market society? Sandel identifies two main problems: "one is about inequality, the other about corruption."[5] The inequality ques-tion is more obvious, and less relevant to our purposes. "In a society where everything is for sale, life is harder for those of modest means."[6] The point about corruption is more subtle. "Markets don't only allocate goods; they express and promote certain attitudes towards the goods being exchanged… some of the good things in life are degraded if turned into commodities."[7]

To take one obvious example, the defence of the realm is accepted as a pub-lic good which is not tradeable or profitable, except in a much broader sense; not just business, but life in a wide sense cannot be carried on satisfactorily if the realm is constantly being threatened or attacked. This is not say that aspects of defence will not be undertaken by private contractors, but that is a separate question.

As regards Earth Observation especially in relation to environmental pro-tection, one could argue that this is at least in part a public good which might fall under an environment department budget or remit. (In practice, unfor-tunately, there is a tendency for environment budgets to suffer cuts, at what-ever detriment to public well-being in the long term). There are of course

---

[4] http://www.theatlantic.com/magazine/archive/2012/04/what-isnt-for-sale/308902/
[5] Ibid.
[6] Ibid.
[7] Ibid.

economic and societal benefits, and more especially the avoidance of costs in relation to unchecked environmental degradation, associated with EO but these tend to be long-term, although less so in terms of agricultural monitoring and management, and disaster relief. Here there is particular relevance in terms of Copernicus governance and the implications of the EC Invitation to Tender offering commercial industry the chance to take over user requirement gathering processes for Sentinels not operated by EUMETSAT. It is at least worth asking whether considering Earth Observation as a "market failure" is not an example of market thinking or practice extending into domains where it is not appropriate.

The Precautionary Principle

Earth Observation and the Copernicus system of environmental monitoring are natural extensions and examples in action of the Precautionary Principle which is enshrined in European law. The Principle derives from the earlier German formulation of Vorsorgeprinzip: this includes the German word for care, Sorge, which is also at the heart of Heidegger's philosophy. Care or Sorge for Heidegger is fundamental to human being-in-the-world.

We might also relate care to stewardship, since responsible stewardship has always implied care. The inclusion of the word care in the German formulation Vorsorgeprinzip gives it very different connotations from its usual English translation Precautionary Principle. Caution and care are not the same thing. Caution might be exercised for purely selfish reasons but care implies a concern for the other.

The Precautionary Principle appears to have been invoked for the first time in the 1982 UN World Charter for Nature. In II 11.b of the Charter, it is stated that "Activities which are likely to pose a significant risk to nature shall be preceded by an exhaustive examination; their proponents shall demonstrate that expected benefits outweigh potential damages to nature, and where potential adverse effects are not fully understood, the activities should not proceed."[8] A slightly less strong version of the Principle appears in Principle 15 of the Rio Declaration: "In order to protect the environment, the precautionary approach shall be widely applied by States according to their capabilities. Where there are threats of serious or irreversible damage, lack of full scientific certainty shall not be use d as a reason for postponing cost-effective measures to prevent environmental degradation."[9]

---

[8] http://www.un.org/documents/ga/res/37/a37r007.htm
[9] http://www.unep.org/documents.multilingual/default.asp?documentid=78&articleid=1163

In a European context, a key text is the February 2, 2000 Commission of the European Communities, Communication from the Commission on the Precautionary Principle. Article 3 of the Summary runs as follows: "The precautionary principle is not defined in the Treaty, which prescribes it only once—to protect the environment. But in practice, its scope is much wider, and specifically where preliminary objective scientific evaluation indicates that there are reasonable grounds for concern that the potentially dangerous effects on the environment, human, animal or plant health may be inconsistent with the high level of protection chosen for the Community."[10]

The Precautionary Principle is "profoundly radical and potentially very unpopular",[11] and throws down multiple challenges. These include challenges to the established scientific method (because the Precautionary Principle argues for action in advance of scientific certainty), to the application of Cost Benefit Analysis, to legal principles and especially the burden of proof, and to politicians, by directing them to think in longer time-frames than short electoral cycles. O'Riordan and Jordan define the Principle as follows: "decision-makers should act in advance of scientific certainty to protect the environment from harm...Critical natural habitats should be left intact."[12]

Further, they identify five themes of the Precautionary Principle. First, it is in favour of taking action in advance of scientific proof, thus resisting the temptation always to argue, as happened in the case of acid rain and damage to the ozone layer, that "further research is needed." Secondly, it is fundamentally concerned with safeguarding ecological space. This make the kind of attack on the Precautionary Principle carried out by Cass Sunstein, who argues that the 2003 invasion of Iraq constituted an example of the Precautionary Principle, seem especially misdirected.[13] The Precautionary Principle was developed and operates almost entirely in environmental terrain and is concerned with safeguarding from harm. To use it in connection with a highly risky and violent military operation which cost hundreds of thousands of lives is perverse. Thirdly, the Principle shifts the burden of proof onto developers and perpetrators. Fourthly, the Principle favours long-term thinking and meso-scale planning, with a time-frame of 25–100 years. Finally, the Principle is ostensibly in favour of cost-effectiveness, however difficult this may be to gauge when, almost by definition, many factors are uncertain and difficult to evaluate in monetary terms.

---

[10] http://eur-lex.europa.eu/legal-content/EN/ALL/?uri=CELEX%3A52000DC0001

[11] Timothy O'Riordan and Andrew Jordan, 'The Precautionary Principle in Contemporary Environmental Politics', *Environmental Values*, vol. 4, no. 3, 1995, pp. 193.

[12] Ibid, p. 194.

[13] See Cass Sunstein, *Laws of Fear: Beyond the Precautionary Principle* (Cambridge: Cambridge University Press, 2005).

Taking these themes one by one, the first is only a challenge to a misguided idea of science which views science as absolute knowledge and certainty. Climate change is an area of science which is particularly prone to uncertainty, because of the large number of variables, the complexity of subsystems and "the non-linear and chaotic nature of the atmosphere and oceans".[14] Here one can distinguish three kinds of uncertainty: first, uncertainty as data unavailability, secondly uncertainty as ignorance, and thirdly uncertainty as indeterminacy, defined by Brian Wynne as the "unbounded complexity of causal chains and open networks."[15]

Climate science will always exist in a context of multiple uncertainties, and even indeterminacy, but clearly this must not be a recipe for inaction, any more than residual uncertainties over the details of links between tobacco and lung cancer (for instance why some heavy smokers contract the disease and others do not) or between CFCs and the damage to the ozone layer should have been used as arguments merely to continue with research and take no action.

The second theme, that the Precautionary Principle is concerned with safeguarding ecological space, may appear self-evident, but is not. Even the idea of ecological space is contentious, as it implies a concern for ecosystems and habitats for non-human species which is not necessarily shared by those we have termed "neo-environmentalists", as well as by Cornucopians, neo-conservatives and others. Here it seems to this writer that Copernicus should nail its colours more firmly to the mast of environmental and ecological sustainability. The role of satellite EO in monitoring ecological health is important and undersung. For example, satellite data is used to monitor and corroborate environmental treaties such as the Ramsar Convention on Wetlands. As Josef Aschbacher of ESA has argued, "rapid advances in satellite technology,…the increase in the number of available sensors taking more frequent measurements and an increased awareness of the need for global environmental observation have progressively introduced space technology to the environmental community."[16] For example, new satellite techniques are being used to monitor rainforest mass.

The third theme, the shifting of the burden of proof, is controversial and of great importance. In general terms, as we have already remarked, in the climate "debate" we seem to have reached an odd position where contrarians can make unfounded claims and the mainstream science community is put on the back foot and expected to justify a consensus agreed by more than 95 % of climate scientists and supported by the overwhelming majority of peer-

---

[14] Judith Curry, 'Reasoning about climate uncertainty', *Climatic Change* (2011) 108:723.
[15] Ibid, p. 724.
[16] http://www.isprs.org/proceedings/2005/isrse/html/papers/1037.pdf

reviewed papers. This appears to many experienced observers, including Roy Gibson, to be back to front: "if you're a sceptic, you can say what you like; if you're for [the consensus], you have to be very careful."[17]

The more general shifting of the burden of proof in environmental matters has profound political, economic and democratic implications. For one thing, by attempting to shift the burden of proof onto, for instance, environmental NGOs, as has happened with fracking in the US (but not in Germany where the burden of proof rests on the shale gas companies[18]), large industrial concerns are behaving like bullies: they are entirely cognisant of their enormously greater financial, economic and political muscle. Small environmental NGOs, let alone individual citizens, are unlikely to have the resources available to prove their point.

The fourth theme to do with long-term thinking and meso-scale planning is of particular relevance to the Copernicus system of environmental monitoring, and especially the area of climate change. The Copernicus system is intrinsically long-term and future-directed in its thinking and applications. Some applications to be sure, such as precision agriculture and disaster relief and management, are relatively short-term and immediate, but overall the system is directed towards the long term. Climate change in particular is generally perceived, at least in the rich countries of the global North, as a future concern, which will have greater impacts on unborn generations than on present ones: this is a questionable assumption, given the already quite dramatic impacts of climate change in the Arctic and in sub-Saharan Africa, which in turn have knock-on effects such as the increased flow of migrants attempting to make their way to Europe, with huge loss of life.

Regarding the final theme of cost-effectiveness, the potential cost savings of environmental monitoring have always been part of the rationale for Copernicus. One estimate is that every euro invested in environmental monitoring will bring ten euros of benefits. The rather bland language of "cost effectiveness" should not mask the fact that EO has the potential to save lives, perhaps many thousands of lives.

Citizen Science and Citizen Involvement

One of the potentials of the Precautionary Principle identified by O'Riordan and Jordan is for a much more active involvement of citizens in both framing and solving scientific questions. According to the former director of the

---

[17] Roy Gibson, conversation with the author, 30 May 2013.
[18] http://energytransition.de/2015/04/german-fracking-law-takes-shape/

EEA Jacqueline McGlade, there is huge and untapped potential for Earth Observation and the Copernicus system in this field.

EO and Copernicus have thus far operated overwhelmingly according to a top-down, expert-oriented paradigm. This is a system designed by experts and controlled, to some extent, by bureaucrats. Most EU citizens (to judge anecdotally) have not heard of Copernicus and have little idea of its scope. Perhaps this is not surprising, given that citizens have hardly been consulted in the whole process. This reflects the old-fashioned 'Royal Society' view of science in which beneficent scientific knowledge is handed down to an ignorant public by accredited experts. As we have seen, this is by no means the only possible view of science.

Against that view, "increasingly scholars, government policy-makers, scientific professionals, and concerned citizens around the world have begun to promote the idea that citizens should be not just informed of policy decisions but also involved in setting agendas and contributing to solving challenges faced by institutional stewards of scientific and technological activities."[19] One step in the right direction is undoubtedly the open data policy adopted by the Commission. The principle on data is as follows: "In line with its data and information policy, the Copernicus programme provides users with free, full and open access to environmental data."[20] But the policy of "open access" begs the question "accessible to whom?". Thus far the emphasis has been on looking at Copernicus in terms of "user groups", assumed to be governmental bodies or industrial entities—or possibly environmental NGO's. This is in part because of the nature of the data and the difficulty of making it more widely available. A change would be think not just in terms of "user groups" but also of participants. Here the new mobile technology of apps begins to become relevant. A certain number of apps have already become available, for instance monitoring local air quality. NASA has launched initiatives in this area, including the International Space Apps Challenge, which attracted a remarkable response.[21] There could be lessons here for Europeans.

According to McGlade, Copernicus can deliver neighbourhood environmental and climate plans as well as national and regional ones. Here is a potentially very important way of addressing the disconnect between citizens and climate change, and wider environmental degradation. Arguably, climate change seems too remote a threat, both spatially and temporally, for most

---

[19] Amy Kaminski, 'What Place for the People? The Role of the Public and NGOs in Space Innovation and Governance', *Yearbook on Space Policy 2014: The Governance of Space* ed. Cenan Al-Ekabi, Blandina Baranes, Peter Hulsroj and Arne Lahcen (Vienna: Springer Wien New York, 2016), p. 223.

[20] http://www.copernicus.eu/main/data-access

[21] Amy Kaminski, op.cit, p. 224.

European citizens to care deeply about it. But climate change is not just a matter of future projections and models but of present and local experience. A great advantage of Copernicus is that it delivers incontestable data rather than models or projections; now the challenge is to make the data even more accessible, so that citizens appreciate its usefulness (not least for their own health and security, but extending beyond that to the wider public and planetary good) and experience a sense of ownership.

Towards an Environmental Democracy

In an earlier section we discussed the theme of environmental ethics, and noted that such a concept was relatively recent (though with certain important antecedents), and at present might represent a noble ideal rather than a tangible reality. Extending ethical considerations beyond our contemporary human brethren to future generations and non-human species is an enormous challenge for humanity, currently mired in endless bloody conflicts, some of extreme ferocity, as well as structural imbalances with historical roots between richer and poorer nations and parts of the world. Likewise extending the idea of democracy so that long-term environmental sustainability is no longer peripheral but central to its raison d'etre, is also an apparently Herculean task.

What is Environmental Democracy?

Democracy is perhaps a term more banded about in a loose way than rigorously thought through in our time, not least in the European context. Democracy may come from a word used by the Ancient Greeks, but the kind of representative democracy generally in force in the West has very little to do with Athenian democracy of the fifth century B.C. In the Assembly (ekklesia) in Athens, free adult male citizens with two years' military service voted directly on matters of great concern. Citizens might also be elected by lot to serve on the Council (boule) of 400 which served as a deliberating chamber.

This kind of democracy is based on direct citizen involvement. Our kind of democracy generally limits direct citizen involvement to voting every four or five years for a choice of representative. The business of government has become too specialised and time-consuming for it to be possible for citizens to conduct it directly, the argument goes. There is also a crucial role, as Burke argued, for representatives to exercise their judgement on our behalf, rather than merely acting as delegates.

This kind of representative democracy both carries dangers and in certain ways does not answer to new changed conditions, especially in relation to the amount of information available to citizens. The main danger is that citizens become increasingly passive, surrendering their power of shaping and enacting policy to a class of professional politicians, who may become increasingly unrepresentative of the populace at large as they are influenced by lobbyists or captured by special interests.

The term environmental democracy can be heard in two different ways. In the first sense it implies the greater involvement of citizens in environmental policy and decisions. This is not without difficulties, especially the problem of nimbyism, where for example local people might object to windfarms or indeed nuclear power stations being sited in their vicinity, and merely wish to pass on the problem to other communities. However, many of the most significant achievements of the environmental movement have been initiated at grass roots level.

One of the most famous is the Love Canal case in New York State, USA. A former canal was used as a dump site for hazardous chemicals in the 1940s and early 1950s by the Hooker Chemical Company, which subsequently sold the land to the City of Niagara Falls; it was then used partly for the building of a school and around 100 homes. From the 1970s there was growing concern among local residents about an increase in miscarriages and birth defects in children, as well as visible manifestations of toxic waste. The issue was taken up by a local reporter on the Niagara Falls Gazette, Michael Brown, and also by a local resident, Lois Gibbs, who founded the Love Canal Homeowners' Association. The case was investigated by the Environmental Protection Agency and was declared by President Carter to be "one of the grimmest discoveries of the modern era." It led to the passing in 1980 of the Comprehensive Environmental Response Compensation and Liability Act (CERCLA), also known as the Superfund.

Such cases led to the development of networks such as the Access Initiative, "the largest network in the world dedicated to ensuring that citizens have the right and ability to influence decisions about the natural resources that sustain their communities." The rationale for the Initiative is as follows: "Access to information, access to public participation, and access to justice (the three "access rights") are practical means of ensuring that decisions by governments consider sustainable development concerns and the interests of the poor.

"An informed and empowered public monitors government and corporate performance, is alert to problems, challenges the conventional wisdom of government or corporate decision-makers, discusses the issues, organizes social and political change, and demands improvements. Where independent courts supply remedy and redress without political interference, the public can better hold decision makers accountable."[22]

These concerns have been reflected at a governmental level in the US by the "right to know" policy initiated by the 1986 Emergency Planning and Community Right-to-Know Act. In the words of the Environmental Protection Agency, "Every American has the right to know the chemicals to which they may be exposed in their daily living. Right-to-know laws provide information about possible chemical exposures."[23] Right to know extends beyond chemical exposure to information about, for instance, food quality, air pollution, water quality and hazardous waste.

In Europe there is no single "right-to-know" policy but "the EU has signed the UN **Aarhus Convention**, which entered into force on 30 October 2001. The Convention gives you **the right to view environmental information held by public authorities** (this is known as 'access to environmental information'). This can include information on the state of the environment, on policies or measures taken, and on the state of human health and safety if this can be affected by the state of the environment."[24]

All of these initiatives and laws give support to the idea of an environmental democracy in which citizens, not just government agencies and industry, play an active role in deciding which environmental issues need addressing, and how they should be addressed. All of this is crucially dependent on access to information, so free and open access to satellite data must strengthen the hand of an environmental democracy in which citizens are not obedient followers but active shapers. The devil may well be in the detail—that is to say in how precisely the vast mass of data is to be made accessible. This is a matter which requires urgent attention from European policy makers.

Environmental Democracy as Green Governance

The second sense in which the words "environmental democracy" can be heard is as a kind of democratic governance which prioritises environmental concerns—that is, which puts environmental sustainability and flourishing not at the periphery but at the centre of its agenda.

---

[22] http://www.wri.org/our-work/project/access-initiative-tai/commissions
[23] http://www3.epa.gov/epahome/r2k.htm
[24] http://ec.europa.eu/environment/basics/benefits-law/right2know/index_en.htm

This may sound like a pious dream, and recent examples such as the change in environmental direction of the Conservative government in the UK following its outright victory in May 2015 do not give great cause for optimism. It may be remembered that shortly after becoming Prime Minister in 2010 at the head of a Coalition administration, David Cameron promised to lead "the greenest government ever...We've got a real opportunity to drive the green economy to have green jobs".[25] Only three years later, the lobby correspondents of the Sun and the Daily Mail newspapers reported "a key Tory source" saying that the Prime Minister was "going round no. 10 saying: 'We have got to get rid of all this green crap.'" This story has not been confirmed, but neither has it been officially denied, by Downing Street.[26] Perhaps more significant is the series of volte-faces on environmental policies performed by the newly elected Conservative government, including withdrawing support for solar and onshore energy, and reneging on an earlier commitment to zero-carbon homes.

The truth is that very few governments have ever put environmental sustainability at the heart of their agenda. One exception is the series of National Environmental Policy Plans (NEPP) implemented in the Netherlands, starting in 1989 and continuing. Article 1.1 of the Strategy of the Third Environmental Policy Plan explains the rationale of the policy. It is worth citing this article at some length in order to highlight its exemplary attention to matters of both intergenerational and intragenerational equity:

Why an Environmental Policy?

The public authorities in the Netherlands are charged with enhancing the well-being and living standards of all its inhabitants both now and in the future. The protection and enhancement of the living environment is an important aspect of this duty, and is enshrined in Article 21 of the Constitution. In seeking sustainable development—the main objective of environmental policy—it becomes apparent that concern for the environment is part of a wider concern aimed at well-being and living standards. The term "sustainable development" was coined by the UN Brundtland Committee to describe a development which satisfies the needs of the present without compromising the ability of future generations to meet their own needs.

---

[25] http://www.theguardian.com/environment/2010/may/14/cameron-wants-greenest-government-ever
[26] http://www.theguardian.com/politics/2013/nov/21/did-david-cameron-tell-aides-to-get-rid-of-all-the-green-crap

A clean environment fulfills many functions for us, and we seek to continue to utilise these functions in a sustainable manner. It provides the water we drink, the air we breathe, and succours the crops we eat. A clean environment allows us to live safely in green surroundings. In short, a clean environment is not an end in itself, but rather an essential prerequisite for ensuring the Netherlands is a fit and attractive country in which to live, work and pursue recreation.

Sustainable development requires not only that the environment is clean and free of pollution and nuisance, but also that good-quality natural resources are available for all, both now and in the future. It is vital that the distribution of and access to natural resources are fair, not only within the Netherlands but also globally. The government sees energy, biodiversity and physical space as the critical resources for present and future human needs. At the global level, water and food resources also need urgent consideration.

In the global perspective, it is imperative that the resources needed to satisfy human needs are carefully husbanded. A sustainable development can only be achieved in the Netherlands in an international context, recognising that the Netherlands forms part of a larger whole in social, economic and ecological terms."

12

Conclusion: Earth Observation and the Anthropocene

In the International Geosphere-Biosphere Programme's Newsletter 41, in 2000, the Nobel-Prize winning Dutch atmospheric chemist Paul Crutzen and Eugene F. Stormer proposed using a new term, the Anthropocene, for a new geological era marked by "the global effects of human activities."[1] Just over a decade later, in 2013, a group of prestigious cultural and scientific organisations, the Haus der Kulturen der Welt in Berlin, the Max Planck Institut für Wissenschaftgeschichte, the Rachel Carson Center for Environment and Society, Munich, and the Institute for Advanced Sustainability Studies in Potsdam came together to launch the Anthropocene Project. Associated with the Project was the Anthropocene Working Group, commissioned by the International Commission on Stratigraphy, whose aim is to develop a proposal for the ratification of the Anthropocene as a new geological epoch. The implication of all this is that the Holocene epoch, marked by relatively stable climate, has ended.

The earth is certainly facing an unprecedented crisis. Never before has human pressure upon the planet had the effect of changing and destabilising the climate. There have been great extinction events before in the history of the earth, but they have never been caused by human impingement onto habitats and global depletion of stocks and resources. Despite the scientific consensus and the courageous lead taken by some politicians, general public awareness of these phenomena and the need for action is wavering—hardly surprising given the many conflicting and sometimes politicised messages

---

[1] http://www.igbp.net/download/18.316f18321323470177580001401/1376383088452/NL41.pdf

© Springer International Publishing Switzerland 2017 **129**
H. Eyres, *Seeing Our Planet Whole: A Cultural and Ethical View of Earth Observation*, DOI 10.1007/978-3-319-40603-9_12

which crowd the public sphere. This gives especial importance to President Obama's statement in August 2015 that climate change is not a matter for the future but is impacting directly on the present, and particularly on the disadvantaged of the earth.

After the stalling of the Kyoto process, and the disappointing 2009 COP19 conference in Copenhagen (where complete disaster was averted by the hasty cobbling together of the Copenhagen Accord), the December 2015 Paris COP21 agreement was hailed in some quarters as a triumph. It certainly achieved a breakthrough in bringing together richer, developing and poorer countries to sign a universal accord to cut greenhouse gas emissions and limit the most pernicious effects of climate change. Perhaps the biggest surprise was the ambitious target of limiting global warming below 2°c and attempting to maintain temperatures at 1.5°c above pre-industrial levels. The agreement by wealthy countries to contribute $100 billion a year by 2020 to help poorer countries transform their economies was certainly welcome.

It might, however, be premature to join the cheering. The devil is in the detail. The really tough decisions remain to be made, as ever, in the future, as binding targets for emissions reduction are replaced by Intended Nationally Defined Contributions (INDCs). The pledges submitted by more than 180 countries before the conference would fall far short of limiting warming to 2°c—according to some estimates they would result in warming of 2.7°c, well above the level considered dangerous. The hopes for COP21 rest, essentially, in the review mechanism to tighten up INDCs every five years. The moment to begin applauding would be when there were tangible signs of decisive action, signalling a clear and essentially irreversible shift away from "business as usual". A test of such a shift could be the response to the fall in the price of oil. So far, such responses as the UK government's decision to abandon the fuel duty escalator and growing sales of SUVs in the US do not appear to point to such a shift.

Despite the intentions announced in Paris, it seems unlikely that the world will avoid at the very least a 2 degree increase in global average surface temperatures. Some climate scientists, such as Kevin Anderson of the Tyndall Centre for Climate Change Research at Manchester University, argue forcefully that a 2 degree rise, widely accepted by bodies such as the IPCC as the upper limit of relative safety for warming, has become a virtual impossibility: "In 2012, with emissions at a historically high level and with economic growth driving emissions still higher, we simply have no precedent for transforming our economies in line with our commitments to avoid dangerous (or even extremely dangerous) climate change."[2]

---

[2] http://www.whatnext.org/resources/Publications/Volume-III/Single-articles/wnv3_andersson_144.pdf

Warming well above 2°c is already occurring in the Arctic region, and in other parts of the globe. According to the Australian climate scientist Tim Flannery, "in late 2014 Dr Thomas Knutson of the US Geophysical Fluids Dynamics Laboratory at Princeton University, New Jersey, and colleagues published an analysis demonstrating that it is virtually impossible that the extreme heat experienced over Australia in 2013 could have occurred without the influence of human-emitted greenhouse gases."[3] Flannery also believes that the Australian Great Barrier Reef, one of the great natural wonders of the world, which extends for 2300 kilometres off the coast of Queensland, is doomed.

The growing instability of the climate and the increase in extreme weather events linked to global warming (acknowledged by insurance companies such as the Munich Re[4]) only increase the need for accurate measurement of changes occurring to the earth's land surface, including ice caps, deserts and forests, oceans and atmosphere. We are collectively navigating dangerous and unpredictable waters and we need the best instruments we can muster.

The Anthropocene and Adaptation

Another way of putting this is that we must accept the need to adapt to climate change, as well as continuing to press for mitigation, in terms of reducing carbon emissions. We are living in a radically changed world, whether we choose to use the term Anthropocene or not, with new kinds of dangers. Some of the most potentially serious of these are connected with sea level rise. Coastal or estuarial cities such as London, New York, Hong Kong, Shanghai, Mumbai, and Dhaka are obviously particularly vulnerable to small rises in sea level, which can be monitored with extreme precision via satellites. London only avoided severe flooding during the exceptionally wet winter of 2013–4 thanks to the Thames Barrier. However the barrier was closed 50 times during this winter period, compared with an average of fewer than six times per year since its opening in 1984, and set against the Environment Agency's guidance that it should be not be closed more than 50 times a year.[5] Apart from flood defences, other adaptation strategies include the targeting of use of scarce resources, precision agriculture and the establishment of wildfire corridors.

---

[3] http://www.theguardian.com/environment/2015/aug/26/bushfires-heatwaves-and-early-deaths-the-climate-is-changing-before-our-eyes

[4] http://www.munichre.com/en/media-relations/publications/press-releases/2015/2015-03-03-press-release/index.html

[5] http://www.bbc.com/news/uk-england-london-26453484

Geoengineering

Beyond adaptation lies most proactive form of response to climate change. This is geoengineering, defined by the University of Oxford's Geoengineering Programme as "the deliberate large-scale intervention in the Earth's natural systems to counteract climatic change." Geoengineering in this sense has two main components: first, the removal of carbon dioxide via a number of different techniques including mixing biochar with soil to create terra preta, carbon capture and storage, and ocean nourishment including iron fertilisation; secondly, solar radiation management, achieved either through deflecting sunlight away from the earth (for instance with giant mirrors) or increasing the albedo effect, again using a variety of techniques including cloud whitening and the use of stratospheric sulphur particles or aluminium oxide particles.

Geoengineering obviously raises a number of very serious questions, first of all about ethics and potential harm, then in relation to governance, and finally in relation to feasibility and cost-effectiveness. The assessment by the IPCC's Fourth Panel Report in 2007 is highly sceptical: "Geo-engineering options, such as ocean fertilization to remove CO_2 directly from the atmosphere, or blocking sunlight by bringing material into the upper atmosphere, remain largely speculative and unproven, and with the risk of unknown side-effects. Reliable cost estimates for these options have not been published."

Before commenting on any particular scheme of geoengineering, it is worth pausing to reflect on the differences between what one might call the ideology of geoengineering and that of Earth Observation. Earth Observation operates under both the Precautionary Principle and what one might call the Hippocratic principle: "first do no harm". Earth Observation is devoted to obtaining the most accurate and sensitive measurements of planetary changes, not to any kind of large-scale intervention in planetary systems. Some might indeed argue that geoengineering could a precautionary ideology—acting in advance of scientific certainty to ward off an environmental threat—but it crucially lacks the Hippocratic element. The effects of large-scale tampering with climatic processes are not just poorly understood, but, given the complexity of the processes, may never be understood.

In any case, if any kind of large-scale geoengineering scheme were ever to be undertaken, Earth Observation would surely be crucial in terms of monitoring and assessing its impacts.

The Anthropocene, Earth Observation and Human Responsibility

The term "Anthropocene" in general refers to human impact on the planetary environment. But if radically new conditions obtain on Earth, does that not also entail radically new human conceptions of human responsibility? We have suggested throughout this study that Earth Observation should be considered not just as one more tool in the armoury of human domination of the planet, but as an essential ethical instrument, allowing us to observe our environment from "the necessary distance" to see it as an intricately interlinked totality.

The Migration Crisis

The first great crisis of the Anthropocene is not so much a natural phenomenon as a human one—the sudden increase in the flow of refugees and migrants from Syria, Iraq, Afghanistan, Eritrea, South Sudan and other African countries. Satellite Earth Observation has turned out to be a vital resource in tracking and on occasions rescuing boatloads of migrants attempting to make hazardous crossings of the Mediterranean Sea: as a bulletin from ESA's Euronews puts it, "The ongoing migrant crisis in the Mediterranean shows how indispensable satellite technology has become to those involved in saving lives at sea."[6] In October 2015 the European border Agency Frontex reported that "More than 370 people were rescued off the Libyan coast this week after their vessels were detected on satellite images taken as part of Frontex's Eurosur Fusion Services."[7]

Earth Observation and the UN Sustainable Development Goals

Seventeen Sustainable Development Goals (SDGs) were formally adopted at a UN conference in September 2015, to replace the Millennium Development Goals. The SDGs include: "13) take urgent action to combat climate change and its impacts; 14) conserve and sustainably use the oceans, seas and marine resources for sustainable development; 15) protect, restore

---

[6] http://www.esa.int/spaceinvideos/Videos/2015/09/ESA_Euronews_Maritime_security
[7] http://frontex.europa.eu/news/frontex-eurosur-services-help-rescue-370-people-off-libyan-coast-MxXy7S

and promote sustainable use of terrestrial ecosystems, sustainably manage forests, combat desertification and halt and reverse land degradation, and halt biodiversity loss."[8] According to the Group on Earth Observations, "Earth Observations—collected at local, national and global levels, and supported by the best science, tools and technologies—can serve critical, insightful roles in monitoring targets, tracking progress, and helping nations make midcourse corrections."[9]

Satellite Earth Observation is only one of a number of tools for monitoring the health of a planet going through unprecedented change. However, its remarkable development—coming out of the great success of meteorological satellites—and scope and powers give it special importance when it comes to monitoring climate and being, as it were, the canary in the mine. Satellites provide more or less continuous global coverage; they can give extraordinarily detailed measurements of such factors as ice thickness (via Cryosat) and forest mass.

The economic, social and ecological benefits of Satellite EO data are manifold, from improvements in planning and precision agriculture, to the monitoring of biodiversity, water and air quality, deforestation and erosion. But in this study we have attempted to argue for something even beyond those. Our newfound ability to observe the Earth from "the necessary distance" brings with it a new kind of cultural and ethical relationship. In some ways this new relationship is also a return to much older ways of thinking. As we have seen, from the emergence of western culture until the dawn of the scientific revolution, the earth was seen as living being or nurturing mother. In non-western cultures also, such concepts as organic unity, non-intervention and stewardship were stressed. In moving forward into the Anthropocene, it appears, we need to recover some of the wisdom of our ancestors while continuing to deploy our most sophisticated twenty-first century technology.

---

[8] http://www.theguardian.com/global-development/2015/jan/19/sustainable-development-goals-united-nations
[9] http://www.earthobservations.org/me_sevent.php?seid=439

Epilogue

Earth Observation is both the culmination of a process initiated millennia ago of knowing, understanding and getting a clearer view of our planetary home, and the fortuitous side-effect of other ambitions. Earth Observation grew out of the more limited project of satellite meteorology; it was given emotional and ethical force by the Earthrise photograph of the rising earth taken from lunar orbit during the Apollo 8 mission, which was an apparently incidental part of the Apollo lunar exploration programme.

For both the astronauts who were involved with taking the photograph and for millions of others who responded to it, the Earthrise photo had huge significance: it was not just the astonishing beauty of the earth but also its fragility which were revealed. Perhaps the overwhelming feeling conveyed was of nostalgia, remembering the Greek roots of that word in the words for homecoming, *nostos*, and pain or grief, *algos*. It took the act of distancing oneself, of being at the necessary distance, to bring home the essential homeliness of the planet. The photograph has been given the credit, with a certain degree of hyperbole, for inaugurating the environmental movement. An astronaut from a later Apollo mission, Edgar D. Mitchell, described his response as follows: "It was a sense of the Earth being in critical condition, a recognition of the massive insanity which had led man into deeper and deeper crises on the planet."[1]

---

[1] Interview with Francine du Plessix Gray, New York Times magazine, August 1974.

© Springer International Publishing Switzerland 2017   **135**
H. Eyres, *Seeing Our Planet Whole: A Cultural and Ethical View of Earth Observation*, DOI 10.1007/978-3-319-40603-9

Arguably, the insanity and the crises have only deepened. In 1968, when the Earthrise photograph was taken, the hole in the ozone layer had yet to be discovered and little was known about the link between anthropogenic greenhouse gas emissions and climate change. The melting of glaciers and polar icecaps, the acidification of the oceans, the destruction of the Amazonian and Indonesian rainforests and the overfishing of the oceans were at a relatively early stage.

Satellite Earth Observation offers us continuous monitoring of these and other threats to the environment, and to human health and security. It is generally presented as, and regarded as, a techno-scientific tool, of great utilitarian benefit. Although true as far as it goes, this presentation or framing is only partial and could be not just limiting but possibly even damaging. It could serve to reinforce the message that the right relationship of humans to the Earth is of ever more efficient exploitation.

Against that we argue that Earth Observation has another potential, of bringing home to all earth dwellers the singularity, beauty and fragility of our planet, of reminding us of our responsibility towards it (which also means towards those with whom we share it now, towards future generations and other species) and alerting us to the damage we may be doing to it and the potentially catastrophic risks attendant upon our negligence.

Earth Observation systems such as Copernicus were designed and created by scientists and engineers, and have tended to be seen as somewhat arcane resources, available and comprehensible only to the few. Free and open data access implies the opening up and democratisation of this process. Increased involvement on the part of artists, philosophers and communicators can help to broaden the terms of discussion, to make the message less narrow and more resonant. But above all citizens are now empowered to be guardians of the planet, and to pressurise politicians and business leaders to think and act responsibly and in the long-term interests of people and a stable, flourishing earth.

CPSIA information can be obtained
at www.ICGtesting.com
Printed in the USA
LVOW01*1422310117

522739LV00019B/345/P

9 783319 406022